Internet of Healthcare Things

Scrivener Publishing
100 Cummings Center, Suite 541J
Beverly, MA 01915-6106

Publishers at Scrivener
Martin Scrivener (martin@scrivenerpublishing.com)
Phillip Carmical (pcarmical@scrivenerpublishing.com)

Internet of Healthcare Things

Machine Learning for Security and Privacy

Edited by
Kavita Sharma, Yogita Gigras, Vishnu Sharma, D. Jude Hemanth
and
Ramesh Chandra Poonia

Scrivener Publishing

WILEY

This edition first published 2022 by John Wiley & Sons, Inc., 111 River Street, Hoboken, NJ 07030, USA and Scrivener Publishing LLC, 100 Cummings Center, Suite 541J, Beverly, MA 01915, USA

For more information about Scrivener publications please visit www.scrivenerpublishing.com.

Wiley Global Headquarters
111 River Street, Hoboken, NJ 07030, USA

For details of our global editorial offices, customer services, and more information about Wiley products visit us at www.wiley.com.

Library of Congress Cataloging-in-Publication Data

ISBN 978-1-119-79176-8

Cover image: Pixabay.Com
Cover design by Russell Richardson

Set in size of 11pt and Minion Pro by Manila Typesetting Company, Makati, Philippines

Printed in the USA

Contents

Preface

In recent years, the use of the Internet of Things (IoT) has been on the rise worldwide, bringing with it new challenges and possibilities along with new cybersecurity risks in the area of IoT-enabled healthcare. These new challenges involve smart connectivity, high security, and confidentiality, generating big data, reducing total data latency between machine-to-machine interfaces, and reducing bandwidth, complexity, and power consumption. In the healthcare sector, the IoT has made it possible for devices to monitor patient's health remotely, unleashing the capability to ensure their health and allowing physicians to deliver unmatched care. These IoT devices are more efficient for identifying disease in less time with more precision and have an absence of network segmentation, inadequate access control of legacy systems, and an increased susceptible surface area that cyber attackers exploit. Therefore, even though the IoT has significantly impacted healthcare costs and improving treatment results, IoT-enabled healthcare devices' greatest common threats are data safety and confidentiality. Since these devices communicate and obtain data in real-time, cybercriminals can break into the system and steal the Personal Health Information (PHI) of both patients and their doctors. Even so, the IoT is certainly improving the healthcare sector by redefining the scope of the devices.

Another significant threat is the combination of numerous network devices, which creates difficulties in implementing the IoT in the healthcare sector. The enormous amount of data produced by these devices can also impede the ability of doctors to identify diseases. So, this book addresses these issues and provides solutions through authentication and authorization mechanisms, blockchain, fog computing, machine learning algorithm, etc. Machine learning-enabled IoT devices deliver the information concealed in data for fast, computerized responses and enhanced decision-making. This information might be used to plan for upcoming patterns, distinguish anomalies, and expand intelligence by audio, image and video analyses. The IoT permits devices to send information to isolate blockchain networks to generate tamper-resistant accounts of collective transactions. Blockchain empowers business associates to access and share IoT data but without the necessity of central management and control.

As the world is entering the fourth industrial revolution-otherwise known as Industry 4.0-the combination of the IoT with other technologies, such as cybersecurity, big data, cloud computing and blockchain, is fundamentally changing the healthcare industry. The development of these fields is critical in healthcare because it improves the quality and efficiency of treatments and improves the patients' health.

Objective of the Book

This book's main objective is to motivate the reader to use telemedicine facilities to monitor patients in remote areas and gather clinical data for further research. To this end, it provides an overview of the Internet of Healthcare Things (IoHT) and discusses the significant threats: the data security and data privacy of health records. Another major threat is the combination of numerous devices and protocols, precision time, data overloading, etc. In the IoHT, multiple devices are connected and communicate through specific protocols. Therefore, the application of emerging technologies to mitigate these threats and provide secure data communication over the network is discussed. This book also discusses the integration of machine learning with the IoHT for analyzing vast amounts of data for predicting diseases more accurately. Case studies are also given to verify the concepts presented in the book.

Organization of the Book

The 12 chapters of the book are organized into three sections. The first section consists of three chapters on data security and privacy concerns in the IoHT. The second section contains five chapters describing the application of machine learning, blockchain, and fog computing in the IoHT. The third section discusses the latest case studies in the healthcare sector.

Section 1: Security and Privacy Concerns in the IoHT (Chapters 1–3)

- Chapter 1 discusses the changes and standards required for the healthcare sector, covering privacy and security concerns, requirements, existing solutions, future challenges, and privacy-preserving methods. This chapter focus to enhance the knowledge of monitoring for adverse medical events and leading to a rise in the quality of treatment for diseases.

- Chapter 2 states that the purpose of the IoHT is to enrich the users' experience by providing a responsive, discernible, seamless information service and denial of service to unauthorized proponents. In terms of security, the protection for IoT systems in the healthcare sector emanates from both physical and virtual access. Devices and equipment engaged in the IoHT will only be made accessible to authorized personnel, and this denial of physical access to strangers would in itself ensure the avoidance of cyber intrusion in a significant number of cases. The next layer of protection is at the virtual level, wherein the identity of the access seeker is authenticated by secured protocols and standard measures deployed in the system's design.
- Chapter 3 discusses security and privacy issues at every stage of healthcare data's big data life cycle. This chapter also provides an overview of different laws applicable to the protection of healthcare data in different countries, such as the Health Insurance Portability and Accountability Act (HIPAA), Data Protection Act (DPA), Personal Information Protection and Electronic Documents Act (PIPEDA), etc. The chapter focuses on finding a reasonable explanation for data protection and security behind crucial health information in the future.

Section 2: Application of Machine Learning, Blockchain and Fog Computing in the IoHT (Chapters 4–8)

- Chapter 4 gives an overview of the Internet of Things (IoT) and its applications in several fields, and the security challenges faced while deploying it. Then the chapter focuses on its application in the healthcare field. The architectural design of IoT-enabled healthcare is illustrated along with the benefits, necessities and several challenges faced when using it. The security protocols that researchers have designed for intrusion detection in the IoHT are discussed, along with the further application of machine learning and its contribution to intrusion detection in the IoHT. Moreover, a direction is given for future research towards designing a secure IoHT framework with low latency and fast processing ability for accurate end-to-end data delivery.

- Chapter 5 presents a sentiment analysis and topic modelling-based approach for early warning of a health crisis, which can be integrated with the IoHT's framework and can be fruitful in assisting medical specialists. In this research, information related to the IoHT is collected, followed by dataset generation. Modelling is performed on the proposed IoHT dataset for predicting trends in IoHT domains. Sentiment analysis towards the IoHT's applicability is evaluated to find the overall sentiment orientation of people. In the current pandemic situation, variations in the sentiment orientation of users towards IoHT systems are evaluated, and analysis is carried out to determine the effectiveness of these systems.
- Chapter 6 focuses on telemedicine, providing medical services and clinical data and administering medicine to patients in the current pandemic. It is a cooperative for communication between specialists, drug specialists, patients, and individuals in remote regions. Telemedicine systems are progressively being used by patients, clinicians, and organizations. This chapter illustrates the few steps, which can consider in the case of emergence.
- Chapter 7 discusses blockchain networks in the healthcare sector, focusing on the multilayer IoT/blockchain grounded on architecture customized and planned to be utilized in the medical field. The role of several parties and health service providers, doctors, insurance companies, and pharmacies are integrated with this work. The decisive goal is to crack the problem of performance and scalability. This chapter addresses the convergence across different elements, such as modern architecture, device designs, process, scheme, paradigm, platform, approach, protocol, and algorithm, upon the mechanism designs of decentralized healthcare implementation. It also discusses suitable security solutions, like lightweight cryptographic procedures and protocols, which are challenged with lowering the overhead in the rankings of computations and resources. This leads to the inference that designing an effective intrusion discovery/prevention system that collaborates with dynamic data processing is required. The chapter concludes with a blockchain-based security solution is proposed that is divided into three distinct layers to distinguish and avoid attacks and authorize patient details when registered in cloud-based applications.

- Chapter 8 reviews healthcare systems and their challenges, followed by blockchain and its integration with healthcare. Blockchain can be securely implemented in healthcare systems for sharing, storing, and creating electronic medical records and prescriptions, personalized medicine, remote monitoring, mobile health applications, and the Internet of Medical Things (IoMT), which are vital to improving the quality of patient care. Fraud detection, insurance claims, medical transactions, checking for counterfeit drugs, and tracking, along with other uses for patients, doctors, and healthcare institutes, can also be achieved by using blockchain in healthcare. This chapter ends with a discussion of the challenges faced when implementing blockchain in healthcare.

Section 3: Case Studies on Healthcare (Chapters 9–12)

- Chapter 9 addresses the dark web. The internet is an open platform for communicating across boundaries. Without the usual security checks, the anonymous part of the internet, called the dark web, is used by cybercriminals to perform illicit activities. Security research agencies are continuously designing and executing covert operations to track criminals. For instance, organ trafficking is sometimes used by patients who urgently need a quick organ transplant, but due to the scarcity and long-term storage deficiency, healthy living donors are required who are trafficked from multiple nations to avoid being tracked by security agencies. Here, the rich pay the cost of the organ to the donor. But the secure communication and exchange occur using intermediate agents who manage safe locations, operating doctors, and the patient's travel and recuperation after the operation. So, this chapter discusses the illicit trafficking market, including other criminal activities done over dark web platforms.
- Chapter 10 provides an overview of deep learning algorithms, neural networks, random forest, and decision tree classifiers for analyzing patients' data to predict heart disease. This chapter shows how the medical practitioner can detect heart disease by attaining precise troponin levels and prescribe effective medicine according to the foreseen disease. The study results presented in this chapter demonstrate that heart disease can be predicted with 90% accuracy.

- Chapter 11 describes how the Internet of Things (IoT) has transformed the routine and lifestyle of individuals and its involvement in the arena of healthcare. It is grounded on different machine learning applications, and information is mined for real-time scrutiny of data and secluded health supervision constructed on IoT infrastructure. It helps in forecasting schemes for using machine learning methods, like MLP, Bayes net, SVM, J48, decision trees, etc., in experimental results for breast cancer. In this chapter, features selection, growing efficiency, and deep neural network classification approaches will be exploited to further boost the investigative procedure's performance for breast cancer diagnosis.
- Chapter 12 aims to enhance the reader's understanding of how the Internet of Things (IoT) is used in various medical treatment areas, such as brain cancer. To be trusted Internet of Healthcare Things (IoHT), machine learning algorithms have settled on a colossal commitment to decision-making, which has been clarified through a portion of the contextual investigations in this chapter. This chapter explains the exact therapy of brain tumors, beginning with the types to the best reasonable investigative techniques and survival rates.

We would like to thank all the authors who kindly contributed their chapters to this book. We are also grateful to the publishing and production teams at Scrivener Publishing Group for their assistance in the preparation and publication of this book.

Dr. Kavita Sharma
Department of CSE, Galgotias College of Engineering & Technology,
Greater Noida, India
Dr. Yogita Gigras
Department of CSE and IT, The NorthCap University,
Gurugram, India
Dr. Vishnu Sharma
Department of CSE, Galgotias College of Engineering & Technology,
Greater Noida, India
Dr. D. Jude Hemanth
Department of ECE, Karunya University,
Coimbatore, India
Dr. Ramesh Chandra Poonia
CHRIST (Deemed to be University),
Bangalore, Karnataka, India

Section 1

SECURITY AND PRIVACY CONCERN IN IoHT

1

Data Security and Privacy Concern in the Healthcare System

Ahuja Sourav

The NorthCap University, Gurugram, India

Abstract
Change is the new convention and standard for the world's health sector. Change in the collection, storage, and transfer of healthcare data through digitization is experiencing a fundamental as well as dramatic move in the executing, business and clinical archetype in the world's economy for the predictable future. This change is propelled by lifestyle changes and mellowing populations, the development of software applications, smart devices and gadgets, progressive treatments, and increased emphasis on care value and quality. All these will offer remarkable opportunities for improving and helping clinical objective decision, hence improving healthcare facilities and delivery, monitoring adverse events, and leading to a rise in the standards of treatment for diseases. In this chapter, we will read about developments in medical systems, requirements in medical data systems, and security and privacy concerns.

***Keywords*:** Healthcare system, security, privacy, eHealth data

1.1 Introduction

Medical data is being considered to improve health conditions, gain precious perceptions, and lower medical charges. However, increased privacy and security concerns are so irresistible that we are not able to take complete benefits of it with its available resources and facilities. The diverse technologies which generate data in medical fields are ever-increasing. The information generated is available at the central location and can be

Email: ahujasourav07@gmail.com

Kavita Sharma, Yogita Gigras, Vishnu Sharma, D. Jude Hemanth and Ramesh Chandra Poonia (eds.) Internet of Healthcare Things: Machine Learning for Security and Privacy, (3–26) © 2022 Scrivener Publishing LLC

utilized for any organization including healthcare medical organizations, pharmaceutical manufacturers, hospitals, health insurance companies and all its sub-sectors [1, 15]. The flood of such large data has never been seen before in the healthcare sector. The success of all healthcare organizations needs to manage the analytical power of big healthcare data. This chapter covers privacy and security concerns, level of threat to information in healthcare organizations, security and privacy requirements in the medical field, existing solutions for security of healthcare data, future challenges and work in security and privacy of healthcare data, and privacy-preserving methods [5, 20].

Medical care is evolving as an existing infrastructure and is getting integrated with new technologies. New technologies including digital medical records and sensor-based monitoring of in-home patient remotely are at the current front row. Conventional paper-based medical reports are being digitized allowing patients to easily obtain their records and empowering healthcare workers to regularly monitor the health of the patients and save and protect their lives at the earliest. Specialized sensors and devices can be used at homes for remote patient monitoring [7, 9].

Patients remote monitoring offers powerful and flexible health monitoring through devices anywhere and at any time. The increasing demand, practicality, and convenience of digital medical care have brought various notable challenges for medical organizations, policymakers, patients, and medical centers [28]. A major challenge is dealing with privacy concerns which include patients' personal medical information which is kept online on third party clouds. It is the most significant and testing issue that medical care organizations must face in the future and preserve the patient's medical data and reduce the concerns of their privacy [6, 10].

Generally, the healthcare IoT devices layout is made up of the below-mentioned layers:

- Perception layer
- Network layer
- Application layer.

The perception layer collects healthcare data with the help of various devices. The network layer is composed of both wired as well as wireless systems and middleware. The network layer's major task is to transfer and transform the input data collected by the perception layer. The transport layer is well-designed. It improves transmission efficiency and reduces energy consumption. The transport layer also ensures privacy and security. The application layer provides healthcare services individually to users and satisfies their needs by integrating medical information resources in accordance with the real situation of the service demand and target users [3, 19].

Data analytics in medical services has many benefits, and it displays extremely high potential for improving the medical care systems, yet it has many hurdles and challenges. Therefore, medical care data privacy and security concerns increase day by day. Also, a reactive, technology-centric method to determine privacy and security needs is not efficient by healthcare organizations to protect the medical centers.

The combination of these technologies will reduce costs and medical errors while enhancing the medical care standards by making them more sophisticated. To make new technologies and advancements in the field of medical sciences socially acceptable, related concerns of security need to be looked upon [24].

Security and privacy concerns of healthcare information are two required and essential concepts. Security of information means that data is collected, stored, and transmitted through secure channels. Data security guarantees validity, authenticity, and integrity. By privacy of data, we mean that the data can be retrieved by the users who have the authorization to access it [3, 18]. Based on different purposes and requirements, more reasonable protection strategies could be developed. The advancements and widespread use of healthcare sensors and devices provide a better guarantee to protect people's health. The advancements in the healthcare and medical field also put data privacy and security preservation under pressure. Security, as well as privacy, is the core consideration in the successful development of medical healthcare devices. New data storage and information systems along with new methods are required to avert breaches and other security happenings of delicate data to make efficient utilization of the large medical care information [11].

Both securities, as well as privacy, have impacted the way companies store, manage, and analyze sensitive data all over the world. Medical sciences along with information collected have enough capability to improve health results, predict the spread of diseases, get knowledgeable viewpoints related to the health of the society, prevent diseases in time, cut down the cost while also improving the standards of medical systems and raise the standards of living. Also, while securing user's privacy right, determining allowable uses of information is a difficult job [26].

To guarantee trustworthy and appropriate healthcare data, it is important to know the shortcomings of available systems and solutions in place. Privacy is typically known as the ability to preserve critical data about individuals related to medical reports. It aims to utilize and govern the private data of the users which includes forming new standards and ensuring consent to ensure that users' data is being stored, transmitted, and used incorrect and secure course of action. Security is often known as preventing an unauthorized entry into the user's data. It also includes the availability and integrity of the data. It aims to protect sensitive information from deleterious attacks and information breach. Security is essential for protecting sensitive information, but it does not address privacy [1, 16].

1.2 Privacy and Security Concerns on E-Health Data

Security and privacy issues emerge from various vulnerabilities of information from the entire medical sector. It can happen between users, providers, and payers. The concerns of security and privacy in the healthcare sector are divided into two categories:

- Security and privacy worries about the improper release of data from individual organizations: Improper release of data from individual organizations can happen from following:
  - authorized users who access or spread data by infringement of organizational rules intentionally or unintentionally
  - The intruder who breaks into an organization's system.
- Privacy and security concerns regarding the systemic flows of data in the medical sector field and related sectors: It includes the open revelation of the patient personal medical data to parties that might be in opposition to the patient, and he might have thought of as breaching and violating the privacy of the patient [2].

1.3 Levels of Threat to Information in Healthcare Organizations

There are different types and levels of distinct types of organizational threats. The following threats are categorized as levels numbered from 1 to 5:

- Threat 1: *Insiders who do silly mistakes unknowingly which causes accidental revelations:* It is likely the most common source of privacy breaching which includes accidental disclosure of personal information. It can happen in many following ways:

- Conversations between healthcare providers might be overheard anywhere
- Test results being noticed by the laboratory in-charge for familiarity in laboratory tests getting processed
- Computer screen left turn on which can lead to the passer seeing the information
- E-mail or fax sent to the wrong address
- Data left misclassified or misfiled.

- Threat 2: *Trustworthy insiders who misuse the privileges and access of data given to them:* These are the people who are given authorized access to medical records and infringe the trust corresponding to the privileges given to them. Medical workers, neither have the need nor the right to know the data but still, they are subject to curiosity in accessing information. There is no overall official statistics to indicate how many healthcare workers have accessed the medical records of others. Some reports indicate that medical care providers have seen medical records of others to estimate the chances of the sexually transmitted disease of others, details about their personal lives, with whom they were in close relation, with whom their former spouses were in relation. Medical staff tend to see the information about celebrities, politicians, actors, sports personalities, and other famous people which is potentially embarrassing health information, and the news regularly spreads out in the news.
- Threat 3: *Trustworthy people of an organization who intentionally investigate the data for some selfish motive:* The threat arises when that insider is authenticated to the partial data but not to the information he desires and through digital forensics and cyber tools gain unapproved access to that information.
- Threat 4: *Physical intruder who access the unauthorized data:* The invader has access to information but does not have authority and permissions for data access and to see the medical records. Many attackers pretend to be an authorized person by putting on a dress code and start using the system and get the desired health information of others.

- Threat 5: *Outsiders and employees who are vengeful who plans attacks to access the information they are unauthorized to, damage hospital assets, and obstruct medical operations:* The attacker does not have authority and physical access but still gain access to the data by purely technical means. Therefore, it is also called a pure technical threat. For example, the attacker breaking into a system from an external network applies cyber attacking tools and retrieves patient sensitive health records. Threat 5 is extremely dangerous. It is prominent that many healthcare organizations are heading towards using computer networks and distributed computing technologies in storing medical records, and so increasing the risk as patient records can be accessed through an external network. This is a dormant problem on the planet. It also has the risk of DOS attacks conducted by outsiders. DOS makes the attacked device useless for normal operations. An example of this can be, an outside attacker may access a sensitive medical information system and place a computer virus or Trojan horse into the computer that crashes the computer or erases important data healthcare records. Also, an intruder can burst your system with thousands of emails and could launch an e-mail attack. He/She can send a large number of emails in a span of less time, which is beyond the capacity of the mailing servers to process all emails. It leaves the computer and leaving the system futile for e-mail purpose [2, 29].

The invasion of privacy of the patient is a burgeoning concern in the field of information analytics. An incident reported quoted that Target Corporation sent baby care coupons to a teenage girl unbeknown to her parents. The incident raises an alarm over patient privacy and impels information analytics to consider privacy. Furthermore, it is essential for driving healthcare analytics to use privacy-protecting cryptographic encryption methods that permit managing algorithms of prediction on ciphertext while also preserving user specifications. However, the resource-constrained environment has the challenge to do the operation of exhausting resources while preserving privacy [1, 17].

For long, medical devices have been facing a lot of difficulties like, writing diagnoses on paper, making it difficult for doctors to access their patients' information. There is also a lack of space, time, and personnel to monitor patients. Development in the systems provide opportunities to better the healthcare systems and minimize the problems. It would also provide personalized service. With the advancement, the privacy and confidentiality of patient's health records are at stake. Internet and network expose their personal information to hostile attacks. Attackers take advantage of the flaws and inadequacies of the system to collect the records of medical data and uncover it in front of the world. The traditional system does not face this problem as patients have to go into a medical laboratory or hospitals to receive their health reports which limit the count of people viewing the medical reports. Medical reports or records, once available online, open illegitimate doors for spammers and hostile attackers to capture and attack the records. Therefore, it is a big challenge to preserve information security as well as authenticity as opposed to the conventional medical care systems. While implementing EPRs as well as networking sensors, storage, authenticity, and access of information are the main challenges.

In the aspect of dealing with the challenges of electronic information, one should be aware of the following questions:

- Who is the owner of the information?
- What and what amount of information to be stored?
- At what place should the information be stored?
- Who all can be viewers of patients' medical records?
- Who can access this information without the patient's consent?

Rapid growth in the medical sector is enabling e-health records like EPRs in making the health reports available via the Internet to users/patients. and empowering the healthcare system being used for proper monitoring and analysis. In addition, the idea of monitoring patients remotely is turning into a practice by advancements in sensor networks [9, 14].

1.4 Security and Privacy Requirement

Although concepts of privacy and security are vital in medical healthcare systems, the maximum organization in medical sector does not spend many resources to preserve these constraints. Healthcare devices produce a highly sensitively large amount of real-time diverse data. In such a scenario, comprising data privacy and security and knockdown the security of the healthcare system or server could cause dangerous outcomes. At each step of information collection, storage, transmission, and analysis, the user's medical data exists. This can cause privacy breaches from multiple points of failure. While developing medical healthcare devices, equipment, and network privacy and security systems, the below-mentioned needs should be taken care of.

- Data Integrity: Data integrity means that information satisfies linguistic standards without tampering through unauthorized and unacceptable methods. It comprises two levels of reliability as well as accuracy. It can be sub-divided into the following four categories:

  - Entity integrity
  - Domain integrity
  - Referential integrity
  - User defined integrity.

  These categories are managed by constraints, foreign keys, and triggers.
- Data Usability: Usability of data ensures the usage of data and data systems by authorized users. A large amount of data brings advantages and censorious challenges like deviant information and dirty data. Moreover, the usability of data is also ruined by data usage and exploitation due to unauthorized access.
- Data Auditing: Healthcare information audit access is a productive method to analyze resources utilization. Any abnormal events in the human body can be tracked by this measure. Also, cagey roles played by cloud service providers require rational auditing ways. Audit information usually includes those service providers, users, operation records, access, etc.
- Patient Information Privacy: Patient Information Privacy is the most crucial requirements in medical data privacy and security. Users' data is divided into below mentioned categories:

  - General patients' records
  - Critical information.

Critical information consists of information like identity information, sexual orientation, mental condition, sexual functioning, fertility status, infectious diseases, genetic information, and drug addiction. With such information about patient medical privacy, healthcare organizations need to ensure that such information is not breached by unauthorized users. These organizations should also use cryptographic encryption to make sure if in case data is breached and intercepted, unauthorized users cannot understand the information and the patient's sensitive data is preserved [3, 30].

1.5 Security of Healthcare Data

Medical organizations collect, save, and manage a large amount of data for smooth functioning of the digital healthcare systems and providing efficient and proper healthcare to patients, but it has its downsides i.e., the shortage of technical support and minimum security. In recent times, healthcare systems are most vulnerable to data loss and exploitation which are extremely complicating. Attackers make use of data mining methods to get access to critical data and takes advantage of the vulnerability and causes critical information breach. The stakes for data privacy and security are continually raised but the ways for establishing and implementing security measures and defeating security controls remain a complex process and become more sophisticated gradually [27].

Therefore, keeping all the above-mentioned points in mind, organizations must establish and implement information security systems and solutions that will help preserve important healthcare assets which satisfy mandates of medical compliance.

1.5.1 Existing Solutions

Since healthcare devices and equipment are not so reliable concerning enough memory storage, processing power, and communication potential, they require an expanding, high-speed, and powerful big infrastructure for storage of information, data analysis and computing [3]. These concerns involving information access, storage, transmission, and processing are not new to the healthcare field [9]. At present, most healthcare organizations utilize cloud services from a third party and store the obtained healthcare records and deploy their applications on the servers. These devices are synchronized with their medical care tasks to the servers accordingly and store data on the cloud. Cloud services facilitate a propitious solution for managing ever-increasing and pervasive medical data by their flexibility and facility to utilize common resources and infrastructure in an omnipresent way [3].

The most widely used technologies to secure the privacy and security of medical data are as follows:

1) Authentication: Authentication confirms or establishes that the claims made about authentication and truthfulness are true. It serves essential functions in any organization like protection of the user identification, ensuring that the user claims of authentication are real and is the same person who is pretending to be an authentic user, and securing access to corporate networks. Data Authentication methods are used to confirm that the information sent is coming from the user it is being claimed to be from. Data authentication can have attacks like man-in-the-middle attacks. Many cryptographic protocols take measures to prevent man-in-the-middle attacks by including some form of endpoint authentication. A man-in-the-middle attack may be affected by persuading a user to agree with a certificate of digitalX 509. Therefore, while designing and building an end-to-end authentication solution, such scenarios must be considered. For instance, for providing communications network security such as over Internet protocols like transport layer security, secure sockets layer are cryptographic protocols that transport layer security and secure sockets layer are in use which encrypt end-to-end the network links segments at the transport layer. Browsing web pages, e-mails, faxing, real-time messaging, and voice-over-internet protocol (VoIP) is also controlled and managed by several versions of the protocols. To achieve authentication, hashing SHA-256 and the Kerberos mechanism can also be used. For authentication of servers, one can use SSL or TLS using a mutually trusted certification authority. To monitor all sensitive information, the algorithm of Bull Eye is useful. The algorithm of Bull Eye manages relationships between actual data and duplicated data and ensure information security. Both medical data obtained from the providers as well as the consumers' identities must be verified before giving access at each entry-level in healthcare systems. Various authentication protocols like passwords, digital signatures have been designed for better energy-efficient sensor networks, like the TinySec hash function [1, 5].

2) Encryption: Data encryption refers to preventing unauthorized access to critical data efficiently. It maintains as well as protects owner possession of information from the data center to the endpoint and into the cloud. It also includes electronic gadgets used by clinicians, administrators, and physicians. Encryption can be used to prevent theft of storage electronic devices and packet sniffing [1]. Cryptographic encryption is a technology for secure communication and data exchange in which the actual message is encrypted by the encryption algorithm into ciphertext by the agreed rules. The message is transferred from the sender of the message to the receiver through the public channel and the message is finally decrypted into plain text for the receiver to read. Encryption can be used to help

prevent eavesdropping and skimming and to make sure information security. Encryption can be done in both software and hardware. Using both software and hardware encryption guarantees the highest-level security.

Medical care organizations must ensure that the encryption algorithms are efficient. Healthcare providers should make sure that the cryptographic encryption method is efficient and easily usable by patients as well as medical professionals. It must also be simply and effectively able to include a large amount of new electronic medical records. Also, it is essential to take care that the number of keys holds by each party involved in the healthcare system should be minimized for easy maintenance and ease of usage. Encryption algorithms like RSA, AES, DES, triple DES, RC4, Rijndael, Blowfish, IDEA, and many more can be used for efficient encryption. It is a tough job to choose properly the appropriate encryption schemes to ensure storage security [1, 14]. Protocols of management play a very key role in the process of securing e-healthcare devices and their communication. But it is also evident that complex transmission protocols or encryption schemes can diversely impact the rate of transmission. It can even sometimes fail to do information transmission. Also, they are required to use valuable healthcare assets that are not readily available. Maintaining a balance between protection of security and the energy utilization of the system is tough and we need to be cautious while solving this as well as remain scientific with our approach.

3) Data masking: Masking replaces critical information with anonymous data. Actual data cannot be obtained from the masked anonymous data as this is not truly an encryption algorithm. It utilizes a de-identifying strategy of information sets which means masking personal identifiers like name, security number, and generalizing or suppressing quasi identifiers like zip codes and date of birth. Therefore, masking data is a famous approach to live unidentifiable information. Data masking reduces the cost of securing a data deployment which adds up to its significant benefit. As secure data, masking decreases the need for using extra measures for security on that data after it is sent from a secure source into the platform and while it is in the platform [1].

4) Access control: Access control method is used to define the identity of a user by the data system. The predetermined strategy and policy prevent unauthorized users to access resources. Once the user is authenticated, he should be allowed to access an information system. But even after authentication, access control policy should still monitor their work depending on the access rights and privileges of every authorized user, if applicable. Role based access control is the supreme structure for advanced authorization control. Complexity and cost reduction of administration in huge

networked applications are the best possible results for the access control mechanism. This mechanism is immensely powerful and pliable to grant accesses to the medical sector. It offers advanced controls and authorization permissions to make sure that users can perform operations like information access, cluster administration, and job submission.

There are a lot of solutions proposed to address the issues of access control and concerns of security. Attribute based access control and role-based access control are the most famous access control models for solutions that are promising to control access control and security operations. In the medical system, attribute-based access control and role based access control have few drawbacks when used alone. We suggest adopting technologies in conjunction to satisfy the needs of access control with other fine-grained security techniques like encryption, and access control methods [1]. The encryption algorithms which can be utilized in access control strategies include:

- Symmetric key encryption
- Asymmetric key encryption
- Attribute-based encryption.

Cryptographic functions rely on keys. The cryptosystem's security is affected by the secret keys' length and its generation strategy. Hence, the life cycle of the security system is determined by the key management mechanism for the cryptosystem. Attribute-based encryption has slowly become a mainstream mechanism due to the ascendable key management and pliable access control policies. Medical data exchange provides electronic medical data with explicit data exchange authorization in an ascertainable manner. Currently available non-cryptographic methods for authentication exhibit several drawbacks. It lacks a reliable and secure strategy for enforcement of access policy in medical information systems. Cryptographic encryption approaches, on the other hand, are too costly, complex, and restricted in identifying policies [3].

5) Monitoring and auditing: Auditing and monitoring security are collecting information to investigate events to find the invaders and intrusions. Monitoring and keeping a record of activities that user does and keeping a log of every access in the medical system in chronological order is called an audit. Cloud servers are not reliable and cannot be trusted completely. The consistency, integrity and availability of healthcare information saved on the cloud might be compromised if information corruption or deletion occurs without the consent of an authorized person. The information guidelines are typically specified by the authorized person for security purposes. This ensures no direct contact of the service provider with the

information. The trusted third party comes up with a high reputation for providing unbiased audit reports. In addition, these trusted third parties can be established to preserve the licit benefits of users on the cloud and enable the accountability of cloud service providers properly. These third-party auditing also has few issues which comprise batch auditing, dynamic auditing, and performance metric auditing. Various auditing methods have been presented in the past few decades which includes logistic regression and support vector machine. These auditing methods can detect suspicious access. However, large-scale promotions are restricted by depending heavily on judgments which are expected and predefined tags. In the present time, unsupervised approaches attract more attention to the healthcare sector [1, 3].

1.5.2 Future Challenges in Security and Privacy in the Healthcare Sector

Developers while developing healthcare devices or programs, to get a better balance in the output, considers various factors and their impact which include security and privacy. For achieving a better security environment, various obstacles need to be addressed and require special attention.

- Insecure Network: Many devices and software rely too much on wireless networks like Wireless Fidelity, as they are convenient and have a lower cost. These systems are supposed to be vulnerable to many attacks, intrusions and penetrations which includes MITM attacks, unauthorized router access, spoofing, brute-force attacks, traffic injections, and DOS attacks. Also, untrusted networking paths that have not been certified possess a greater challenge than most free wireless networks in a public place.
- Lightweight Protocols for Devices: specific policy and proxy rules should be followed by the lightweight and cheaper devices and software/hardware sensor-based applications to deliver services. In the current situation, we must apply comparatively expensive solutions if we wish to deliver higher grade security to the sensors. It is a conflict in the medical healthcare system. The main task of security protection is to develop different levels of security protocols like lightweight security protocols according to application scenarios in the future.
- Data Sharing: Information Island phenomena is a big concern, despite the rapid advancement of medical healthcare

data systems. It is difficult to manage the healthcare data gathered as the level of the information collected from devices of different companies vary diversely. However, the inevitable trend of the future constitutes data collaboration and sharing within heterogeneous systems of medical care. Medical care systems determine the privatization of patient data. Privacy and security enhancement can be done with a hierarchical security model by collaborating general information policies to amalgamate different information which might provide better comprehensible data and can improve the privacy and security of the system [3].

1.5.3 Future Work to be Done in Security and Privacy in the Healthcare Sector

Though there are techniques that are used to cure the privacy and security of healthcare information with the above-mentioned technologies, there is always room for improvement.

- Defining proper attributes for role-based accesses: To put the healthcare system in place, logical attributes need to be established and implemented properly for the role based access. It can either be static rules or dynamic rules based upon what is appropriate as per the requirements.
- Policy development: old policy should be alerted and updated policy should be formulated that can cope with jurisdiction across state. HIPAA rules of privacy are equipped with some groundwork, but better-refined policies should be created to create rules that are clear and also users can bank upon. The step towards EPRs (Electronic Patient Records) and the growing amount of healthcare information will be generated remotely because networking sensors create the ability and need to transfer huge amounts of information rapidly. This entails a complete set of rules and procedures that safeguard a user's security as well as privacy. Many patients are ill-informed or are unsure regarding their privacy rights related to healthcare information. In the current setting as more healthcare information becomes available online and can be transferred easily, the increase in healthcare data will also

broaden the uncertainty of users many folds unless comprehensive guidelines and policies are well determined and well informed [22, 25].

- Rules on patients' privacy at home: Guidelines and policies are required to be formulated that will be regulating what sensor information collected necessitate, where it is stored and who will have control over it. Policies are required to be formed which comprises that the user should have complete control of the information which is stored on a central monitoring storage station or patients should at most have fragmented control. Standards and guidelines should also be established over access controls available to patients for handling their health data and reports.
- Data mining rules and technological measures: Rules, regulations and technological measures of data mining includes which person has the authority to analyze what type of information and the rules on collected anonymous information. If some form of automatic technology is feasible, the appropriate technological methods for ensuring these controls are intact [9].

Security and privacy in medical sciences is a complex undertaking. To summarize the overall problem of security in medical sciences, the following points can be considered:

- An extensive surface area for an invasion
- The cloud services transformation
- The need for data privacy
- The shortage of talent compounds.

In the near future, we should plan to address these issues through innovation and advanced technologies in healthcare that can offer a consistent way to work with electronic healthcare data and to establish policies by supporting a better platform. We should even utilize machine learning algorithms to easily detect malware, Trojans, and other cyberattacks. These technologies can be ideal for medical science to grow and will provide healthcare organizations long-term durable firmness across the security spectrum and privacy needs. Invader detecting and preventing methods are tricky to have on the complete traffic of the network. To resolve this issue, a security surveillance infrastructure is built which will examine DNS network traffic, honeypot data, IP flow data, and HTTP network traffic. The optimal solution is to use data correlation schemes which include storing and processing data in distributed sources. At this stage, to recognize whether the domain name, the data packet or network flow is spiteful, three likelihood metrics have been calculated. An alert pop up in the detection system depending on the score obtained by this calculation, or the process terminates by the prevention system [1].

1.6 Privacy-Preserving Methods in Data

Newer privacy preservation methods are helpful to secure the privacy of the data. They are listed below as follows:

1. De-identification: It is a conventional approach to forbid the revealing of sensitive private data by declining the data that could recognize the patient by the following:
   - the removal of patient's identifiers
   - statistical method where it is verified that patient's enough identifiers have been deleted.

An external may still obtain information help for de-identification. Therefore, de-identification does not guarantee the protection of data-privacy. We can develop more efficient privacy-preserving algorithms which shall be feasible to weaken the threat of re-identification. Few concepts have been introduced to enhance this traditional technique like k-anonymity, t-closeness, and l-diversity.

- k-anonymity: This approach works on the principle that the k value is inversely proportional to the re-identification probability i.e., the more the k, the lesser will be the re-identification probability which can result in deformations of data leading to great information leakage due to k-anonymization. Also, excessive anonymization should not be done as it reduces the usefulness of the disclosed data for the recipients as some of the investigations may produce partial and untrue results or sometimes even becomes impossible. In k-anonymization, the sensitive identifier like a disease will be revealed. The two methods to accomplish k-anonymity for a 'k' are Suppression and Generalization.

  - Suppression: This method includes an '*' which could replace some attributes' values. Some or all the values of a column can be supplanted by '*'.
  - Generalization: This method includes replacing the individual values of the attribute with a broader category.

- L-diversity: we moved towards the l-diversity strategy of data anonymization to reflect upon the actual usefulness of the data. It is one of the batch-based anonymizations. It is used to protect data sets privacy by minimizing data representation granularities. This model has the features of Recursive, Entropy, and Distinct. It is the extended version of the k-anonymity. K-anonymity minimizes the granularities of information display that any data maps onto at the minimum k different data records by utilizing different methods which includes suppression and generalization. The k-anonymity model had few weaknesses that safeguarded identities 'k' individuals are not equal to the respective critical data values that were suppressed or generalized. This weakness is handled by the l-diversity model. This method depends upon the range of sensitive attributes which is the drawback of this method. If want to make data L-diverse fictitious data to be inserted even though delicate accreditation has not as much as different attributes. This counterfeit data may result in problems amid analysis, but it will help improve security. This results in the l-diversity technique depending on unbalanced and similitude strike and hence cannot avoid divulgence of the element.
- T-closeness: This can be considered as further advancement and enhancement of l-diversity batch-based anonymization. Its imitation has the features of equal and hierarchical distance. It extends the l-diversity model by considering the arrangement of data values for the particular aspect and treating the values of an attribute distinctly. The major benefit of the t-closeness approach is that it seizes disclosure of attribute. The major drawback is that the re-identification probability rises as the diversity of data rises.

2. HybrEx: The hybrid execution imitation is a model in cloud computing that deals with concealment and solitude. It uses public clouds only in the following cases:

- Organization's insensitive data
- Computation grouped as public.

Computation grouped as the public is when the organizational body announces that solitude and concealment risk is nil in exporting the information and computing it by utilizing public clouds. But the model executes their private cloud for an organizational body's private, delicate data and calculation. Moreover, the solicitation itself also gets divided and performs in both public as well as private clouds when a supplication requires access to both public as well as private clouds. It provides integration with safety and considers data susceptibility before a job's implementation. The drawback of HybridEx is, it does not manage the key that is executed in the map phase at private and public clouds. Another disadvantage of it is, it works with the cloud only as an opponent [1, 10].

There are four categories in which HybrExMapReduce utilizes both public and private clouds and enables new kinds of applications. The following are the four categories of HybrExMapReduce:

- Map hybrid: The diminished phase is implemented in only one of the clouds while the map phase is implemented in both private as well as public clouds.
- Vertical partitioning: Maps and minimizes tasks that are implemented using public data as the input in the public cloud and save the result in the public cloud after shuffling intermediary data amongst them. A similar task is executed with private data in the private cloud that the works are prepared in segregation.
- Horizontal partitioning: In this, the reduce phase is performed in a private cloud when the map phase is performed only in public clouds.
- Hybrid: The map phase and the diminished phase are executed on both private and public clouds. It is also possible to do data transmission among the clouds.

3. Identity-based anonymization: Identity-based anonymizati tion is a kind of data sanitization. The purpose of identity-based anonymization is privacy protection. It is the process in which the persons described by the data remain anonymous either by cryptographic encryption or eliminating personally identifiable data. The main drawback of this technique is that while

protecting the identities, it involves privacy protection combining anonymization and large data approach to examine usage data [1].

Medical Healthcare equipment is the collection of Internet of things devices coupled to the information servers, to execute the services and processes that expand and expand health protection. Heath care IoT based equipment has come out as an upcoming approach for electronic medical care that stores essential vitals of the patient's body, stores them, keep a check on their pathological details with small gadgets that are easily wearable or implantable sensors, and analyzes the patient's condition to stay pre-active in critical conditions. Advancements in healthcare have proved great prospect in improving the medical care and lifestyle of human beings and support many gadgets from embedded healthcare instruments to the wireless body area network [3, 12]. The digitization of healthcare systems and medical records has set a new wave and paradigm shift in the healthcare industry in recent times. Thus, there is a sheer rise in the amount of data in the healthcare industry which has increased complications, variety, and timeliness. The terminology "big data" has been in a lot of use in digital healthcare units which indicates the cluster of complex and big sets of information, which is far more superior to existing computing, storing data, and communicating capabilities of already existing methods or systems. In the previous twenty years, the hiked cost is at a frightening rate. Estimated medical care expenditure is now at 17.6% of GDP. Here, big data emerges as a possible solution as medical care officials strive for every feasible way to reduce the costs while enhancing the healthcare system, its delivery, and functioning. However, there is a makeable increase in privacy and security matters with the acquiring of big data in medical care systems. At the outset, data centers are made with different levels of safety where patients' Information. Also, most of the information centers which store healthcare data in large volumes have HIPPA (Health Insurance Portability and Accountability Act) certification which aims at ensuring safety policies and procedures. HIPAA certification does not guarantee patient record safety [4, 21].

There is also an increase in the popularity of healthcare cloud solutions which has increased the complexities in safeguarding big, distributed SaaS solutions. There is an invasion of huge data sets with varying data sources and formats that adds an extra burden on data storage and processing. In the burgeoning healthcare industry, examining security threats and foreseeing risk sources in an actual environment is of maximum need. The flood of attacks ranging from stealthy malware to Distributed DoS is a matter of concern in the healthcare industry. It is hard to foresee without taking human cognitive behavior into account which is affected by the rise in social engineering attacks and the risks associated with it [4, 8]. In today's IoT based environment, it is incredibly challenging to implement security and privacy in resource-constrained networks and the complexity will keep expanding with the rapid expansion in the count of Internet of Things-based devices. Typical symmetric and asymmetric distribution of cryptographic keys cannot be expanded to more than a billion Internet of Things-based devices. Therefore, the new solutions key management between disparate networks like IP networks and the Internet of Things were scalable and crucial for the integration of the Internet of Things and big data in a cloud-based environment. It will lead to the seamless interoperability of data of healthcare systems [13, 23].

As the healthcare industry leverages emerging big data, analytics of security analytics will be at the utmost priority of any healthcare industry related design leveraging on expanding technologies in the context of big data. Big data along with SaaS solution will make better knowledgeable decisions hosting Protected Health Information (PHI). Moreover, real-time data analytics can be a challenge in the administration of such systems [4]. If security and privacy concerns are intact, medical IT experts can detect risks on critical patients immediately and take pre-emptive steps to control the deteriorating health condition of an individual [18].

1.7 Conclusion

Big healthcare data can be useful for the advancement of medical science. But in driving such technologies, healthcare, security, and patient privacy are paramount. Managing organizations are not willing to share massive their vast medical records for centralized analytics. Securing patient data management is inevitable when medical cloud networks link and generate large amounts of data [4]. It is especially important for the successful digitization of health systems, but it can be utilized only if concerns of privacy, as well as security, are addressed. As it happens, medical care clouds with huge healthcare information will become notable. Additionally, immediate processing will impel proactive healthcare in a secure and privacy-preserving environment [4]. In recent times, with the increased popularity of healthcare innovations and analytics, new privacy laws need to be put in place to demonstrate the entire process involved in analyzing large medical data and to protect patient privacy. Before performing such analytics and applying algorithms on health information, "informed consent" from patients is duly required [20].

At present, to increase the generation of the huge amount of information as well as to improve the quality of medical services, different medical healthcare devices and software are being developed. The importance of this huge amount of medical care information is self-evident. In future related research, the fact that how to effectively protect data security, privacy and prevent data breaches will occupy an important position at all stages of data flow [3, 18]. Attention has been given to medical healthcare systems. The policies, standards and technical guidelines are still under progress in the healthcare system intending to protect lives at the earliest.

References

1. Abouelmehdi, K., Beni-Hessane, A., Khaloufi, H., Big healthcare data: Preserving security and privacy. *J. Big Data*, 5, 1, 1, 2018.
2. National Research Council, Privacy and Security Concerns Regarding Electronic Health Information, in: *For the Record Protecting Electronic Health Information*, National Academies Press (US), 1997.
3. Sun, W., Cai, Z., Li, Y., Liu, F., Fang, S., Wang, G., Security and privacy in the medical internet of things: A review. *Secur. Commun. Netw.*, 1–8, 2018.
4. Patil, H.K. and Seshadri, R., Big data security and privacy issues in healthcare. *2014 IEEE International Congress on Big Data*, IEEE, 2014.
5. Abouelmehdi, K. *et al.*, Big data security and privacy in healthcare: A Review. *Proc. Comput. Sci.*, 113, 73–80, 2017.
6. Ren, Y., Werner, R., Pazzi, N., Boukerche, A., Monitoring patients via a secure and mobile healthcare system. *IEEE Wireless Commun.*, *17*, 1, 59–65, 2010.
7. Meingast, M., Roosta, T., Sastry, S., Security and Privacy Issues with Healthcare Information Technology. *2006 International Conference of the IEEE Engineering in Medicine and Biology Society*, New York, NY, pp. 5453–5458, 2006.
8. Rodrigues, J.J.P.C. *et al.*, Analysis of the security and privacy requirements of cloud-based electronic health records systems. *J. Med. Internet Res.*, 15, 8, e186, 2013.
9. Kumar, J.S. and Patel, D.R., A survey on internet of things: Security and privacy issues. *Int. J. Comput. App.*, 90, 11, 20, 24–25, 2014.
10. Chen, D. and Zhao, H., Data Security and Privacy Protection Issues in Cloud Computing. *2012 International Conference on Computer Science and Electronics Engineering*, Hangzhou, pp. 647–651, 2012.
11. Wilkowska, W. and Ziefle, M., Privacy and data security in E-health: Requirements from the user's perspective. *Health Inf. J.*, 18, 3, 191–201, 2012.

12. Al Ameen, M., Liu, J., Kwak, K., Security and privacy issues in wireless sensor networks for healthcare applications. *J. Med. Syst.*, 36, 1, 93–101, 2012.
13. Zhang, Y., Zheng, D., Deng, R.H., Security and Privacy in Smart Health: Efficient Policy-Hiding Attribute-Based Access Control. *IEEE Internet Things J.*, 5, 3, 2130–2145, 2018.
14. Rahim, F.A., Ismail, Z., Samy, G.N., Information privacy concerns in electronic healthcare records: A systematic literature review. *2013 International Conference on Research and Innovation in Information Systems (ICRIIS)*, Kuala Lumpur, pp. 504–509, 2013.
15. Burghard, C., *Big data and analytics key to accountable care success*, pp. 1–9, IDC Health Insights, Framingham, 2012.
16. Fernandes, L.M., O'Connor, M., Weaver, V., Big data, bigger outcomes. *J. AHIMA*, 83, 10, 38–43, 2012.
17. van der Linden, H. *et al.*, Inter-organizational future proof EHR systems: A review of the security and privacy related issues. *Int. J. Med. Inform.*, 78, 3, 141–160, 2009.
18. Gong, T., Huang, H., Li, P., Zhang, K., Jiang, H., A Medical Healthcare System for Privacy Protection Based on IoT, in: *Proceedings of the 7th International Symposium on Parallel Architectures, Algorithms, and Programming, PAAP '15*, pp. 217–222, 2015.
19. Hu, J.-X., Chen, C.-L., Fan, C.-L., Wang, K.-H., An intelligent and secure health monitoring scheme using IoT sensor based on cloud computing. *J. Sensors*, 2017, Article ID 3734764, 11 pages, 2017.
20. Groves, P., Kayyali, B., Knott, D., Kuiken, S.V., *The 'big Data' Revolution in Healthcare*, McKinsey & Company, New Jersey, New York, 2013.
21. Lee, W.-B. and Lee, C.-D., A Cryptographic Key Management Solution for HIPAA Privacy/Security Regulations. *IEEE Trans. Info. Tech. Biomed.*, 12, 34–41, 2008.
22. Meingast, M., Roosta, T., Sastry, S., Security and privacy issues with healthcare information technology. *2006 International Conference of the IEEE Engineering in Medicine and Biology Society*, IEEE, 2006.
23. Jiang, D. and ShiWei, C., A Study of Information Security for M2M of IoT. *3rd International Conference on Advanced Computer Theory and Engineering (ICACTE)*, pp. 576–579, 2010.
24. Meingast, M., Roosta, T., Sastry, S., Security and privacy issues with healthcare information technology. *Conf. Proc. IEEE Eng. Med. Biol. Soc. 2006*, vol. 1, pp. 5453–5458, 2006.
25. Choi, Y.B., Capitan, K.E., Krause, J.S., Streeper, M.M., Challenges associated with privacy in healthcare industry: Implementation of HIPAA and the security rules. *J. Med. Syst.*, 30, 57–64, 2006.
26. Corritore, C.L., Kracher, B., Wiedenback, S., Marble, R., Foundations for trust for e-Health, in: *Emerging Healthcare Systems*, M. Ziefle and C. Röcker (Eds.), pp. 49–75, IGI Global, Hershey, PA, 2013.

27. Kargl, F., Lawrence, E., Fischer, M., Lim, Y.Y., Security, privacy and legal issues in pervasive ehealth monitoring systems. *7th International Conference on Mobile Business icmb*, vol. pp. 296–304, 2008.
28. Milenkovic, A., Otto, C., Jovanov, E., Wireless sensor network for personal health monitoring: issues and an implementation. *Comput. Commun.*, 29, 2521–2533, 2006.
29. Samy, G.N., Ahmad, R., Ismail, Z., Threats to Health Information Security, in: *2009 Fifth International Conference on Information Assurance and Security*, pp. 540–543, 2009.
30. Parks, R., Chu, C.-H., Xu, H., Healthcare Information Privacy Research: Issues, Gaps and What Next?, in: *Americas Conference on Information Systems (AMCIS) 2011 Proceedings*, 2011.

2

Authentication and Authorization Mechanisms for Internet of Healthcare Things

Srinivasan Lakshmi Narasimhan

UrbanFirst Systems Private Limited, Chennai, India

Abstract

The potential risk posed by a vulnerable and security-compromised healthcare system, which deploys devices working with the Internet of Things (IoT), weighs extremely high on its consequential damages to its stakeholders – whether they be patients, doctors, other medical professionals, administrators, regulators, insurers or vendors. The risk of harm far outweighs the benefits of unbridled proliferation and thus requires comprehensive resolution. Cyber-attacks could lead to false claims being preferred on insurance companies with stolen patient personal data, fake data could be introduced with hard to measure consequences and botnets could infiltrate the network and cause havoc through DNS attacks or introduction of malware—all of this causes mayhem inpatient healthcare systems. The purpose of introducing innovative applications like the internet of things is to enrich healthcare services, with enhanced quality in speed and delivery of a discernible range of services to the stakeholders, and the same stability is threatened when there is an attack using stolen data. The key to overcoming the challenge of intrusion into the system is to identify genuine users by implementing a robust and comprehensive mechanism of authentication (knowing and validating) and then authorization (allowing access to permissible activities). With the multitude of devices connected to the healthcare system in the IoT scenario, it is rather a formidable and daunting task; however, it is both doable and effective in neutralizing the challenges of hacking. Graded adoption of authentication and authorization is recommended, contingent on the type of use – routine data capture and recording vis-à-vis online procedure performed with or without remote support.

Email: slnarasimhan@urbfirst.com

Kavita Sharma, Yogita Gigras, Vishnu Sharma, D. Jude Hemanth and Ramesh Chandra Poonia (eds.) Internet of Healthcare Things: Machine Learning for Security and Privacy, (27–40) © 2022 Scrivener Publishing LLC

Keywords: Vulnerability, data-security, privacy, Internet of Things, healthcare, devices, sensors and security protocols

2.1 Introduction

The main priority for healthcare organizations providing automated services beginning at registration, up to treatment and follow up, is to protect service beneficiaries and their data. This entails focused contribution from a range of development community supplying devices, applications and system integrators factoring issues of security and privacy inbuilt into their design specially for the domain of IoT in healthcare. Intending to enable and provide optimal protection, one needs to understand the types of threats to which the network and its subsystems and individual components (devices) could be exposed.

Data from the Zion Market Research indicates that the value of the global IoT healthcare market is anticipated to reach $14 billion by 2024 with a CAGR of 12.2% between 2017 and 2024. The increase is attributed to the increasing use of cloud computing in healthcare facilities. One of the principal drivers for the exponential growth envisaged in IoHT is the fast adoption of cloud computing and the accepted norm of storing data and hosting services through the cloud rather than the traditional practice of premises bound.

A survey conducted by McKinsey in 2015 [1], it was found that breach of security scored extremely high on the list of concerns that could potentially impede the growth of IoT with its resultant lack of confidence in the infrastructure and services. In the case of the IoT, exposure to vulnerabilities arising out of low security may be the deterrent—a point underscored by the survey conducted in collaboration with the Global Semiconductor Alliance (GSA). Any discussion on the potential growth opportunities in IoT is weighed down when dealing with the issue of provisioning adequate security and the costs thereof.

Healthcare devices and applications collect, store and pass on personal healthcare data to provide the designed services. In the case of internet, such devices, gadgets and wearables form the nodes of the network and are connected in such a way that they are both accessible and accessed anytime and anywhere as per connectivity requirements. Like the internet, such networks which have their terminals or origins on the internet are prone to hacking and attacks. It is thus necessary and important that the system designers and developers consider and factor in protection elements and considerations while crafting the infrastructure and services to render IoT in the healthcare sector [2].

It is not uncommon to notice that the general requirements of security for any IT based system are equally applicable to the IoT environment; however, the scale and scope of the requirements stand enhanced due to the likely damage that could result in case of any compromise. Preserving confidentiality, including denial of data, by making them inaccessible, whether it is sensitive medical data, or the details of the patients, doctors, providers and supporting institutions to unauthorized users improves the level of security. Integrity and encryption ensure that data, whether it is patient-related or other sensitive data, and is not changed or modified while in transit by any adversary action. More importantly, the integrity of stored data and content should never be compromised, claiming extra vigil and protection. Herein, though the emphasis is on medical data, protection is equally applicable for other forms of data as well as its metadata.

The weakest node or link represents the most vulnerable point prone to attract the attention of attackers, and such weakest nodes either in isolation or on a cumulative basis influence the overall security of the system. It is observed that the most vulnerable, and least resistance-offering category of nodes, are devices that have IoT capability—yet are nevertheless part of the end user device family. The vulnerability arises both from the hardware features of the devices with embedded IoT capabilities, as well as user behavior, determining the willingness to follow and implement recommended security protocols.

Protection in the form of authentication and authorization should be implemented at all devices, interfaces and links to ensure protection against attacks. Both of the protection mechanisms are needed when the device contacts the data transmitter, and when the latter interacts with the server designated to collect and analyze all the data (data aggregator). Entire infrastructure must be equipped with resilience in the form of firewalls or denial of service mechanisms at the data transfer level, to prevent any access to seekers/hackers other than the domain of authorized users.

2.2 Stakeholders in IoHT

The principal stakeholder is the patient, who requires examination and treatment, whether admitted as an in-patient or treated as an out-patient with periodic investigations and medications. In the connected world of today, patients schedule their appointments and consultations by accessing the system of medical service provider through their devices: either by email, messaging, or custom-built apps. Educating, training, and making patients aware of the ways to use the IoHT apps, safely and securely, assumes a significant step toward their data protection, given that the devices and the apps at the edge are the most vulnerable elements in the value chain of IoHT.

Vendors' group that develops applications to provide a range of medical services is the next level of stakeholder. This group should follow standards and protocols for communication, aligned with best industry practices, enabling secure functioning of the individual components as well as the sub-systems. Tendencies and practices that resort to adopting unlicensed and non-standard platforms, and overriding controls, should be avoided. It is the function and responsibility of the system integrator to establish qualification criteria for the selection of applications and devices that conform to, and comply with, secure list of standards and protocols. The often-observed painful practice among the developers is the claim that implementation and incorporation of security features within the applications would not be cost-effective, and the resources required to keep track of the updates and version changes is intensive. The cost-effectiveness can be justified by the damage that could be done to the system; and to ease updates, artificial intelligence may be deployed for scanning and making required changes.

Doctors and other medical professionals form the second important category of stakeholders in IoHT. Doctors are both users as well as knowledge providers, who help craft the services' list based on which applications are developed. Doctors and their associated medical colleagues are the most critical resources in the healthcare value chain, yet present themselves as potential vulnerabilities from a user's perspective. Their behavior and knowledge of operating the IoHT devices is key to the successful protection of data and its security thereof. Importantly, their input and feedback are of paramount importance to segregate data into three categories: sensitive data, personal data and non-personal data. Their judgement of response times and sense of assigning criticality can help design the appropriate security features of the IoHT.

Medical administrators usually make the call to authorize or restrain the budget for a system's architecture, configuration, and security features. The tendency observed among such stakeholders is that cost saving is the driving force behind their measures: their consideration and decisions for system scoping and expenditure is guided by financial considerations. It is thus critical for this group of stakeholders to be made aware of the significance of incorporating security mechanisms right from design to implementation, to ensure the long-term sustainability of the IoHT.

Policy makers, including governments, function as stakeholders who could prime the IoHT landscape with appropriate rules and regulations.

If their policy and regulations require the IoHT sector to comply with security mechanisms to protect privacy and continuity of services, that would greatly support the IoHT.

2.3 IoHT Process Flow

In a typical use case of diagnostics in the healthcare system, data from the patient is collected through devices incorporating sensors, and processed by special and specific applications in a range of equipment like a normal computer or custom-built terminals. Such user terminals are connected to a gateway through short coverage communication protocols, usually conforming to industry adopted IEEE 802.15.4 standard [3]. This gateway connects to a medical server or cloud services for data processing and storage—critical functions in the system operation. There is a choice of segregating the storage of data—either centrally or in a distributed manner depending upon the system architecture and configuration. Segregation and decentralization enhance the level of security for the data; however, that could bring down the efficiency of operation and response time to data handling. Simulated testing of scenarios of data storage and handling is a recommended important step in system design and integration as well.

IoHT can support several use cases in the healthcare environment—patient registration, scheduling appointments, clinical records in the form of noting, images, and reports, recommended and administered medicines, tracking the prognosis of health conditions and follow up procedures. For each of these use cases, custom-built devices and applications are available and it is the task and responsibility of the system integrator to put the pieces together following standards and specifications for optimal functioning of the system. It is at this stage that the consideration of security and privacy issues need special mention in that they require seamless absorption at the design phase.

Unlike other internet of thing's domains, healthcare represents a unique opportunity as well as a challenge since the data in IoHT requires the highest standard of protection in storage and handling. It is not the patients' data alone that is critical, but the entire value chain of data relating to the medical community, financial support providers and that of vendors. IoHT is perhaps the only sector where real time intervention is so essential and valued that it is not acceptable to entertain a situation of inappropriate or doctored data. The selection of industry approved protocols coupled with high-quality hardware is a part of tested means to ensure high performance and assured security. The choice between proprietary and open-source

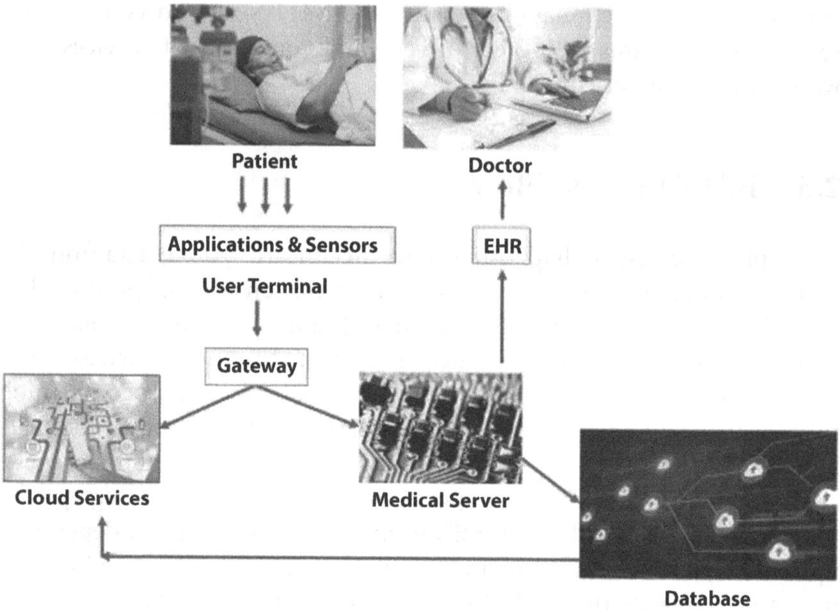

Figure 2.1 Process flow in IoHT. Depicting the flow of data and interfaces (origination and destination of arrows) that are typically vulnerable for cyberattacks. EHR, electronic health record. Image attributions: freepik.com.

protocols and applications is a tricky one and IoHT is not special to escape the exercise of any hard choice. Figure 2.1 illustrates the key interface vulnerabilities that are prone to cyber-attacks. Irrespective of the choices made in the design and selection of devices, it is ultimately the customer perception that would infuse confidence in terms of security—demanding a clear translation of the design of security into demonstrable and vivid results.

With the exploding number of new devices introduced in the healthcare market every day, the designers need to keep the flexibility to scale up performance by incorporating modularity and scalability both in design and implementation.

Access to the medical server and database should be strictly restricted to a special, authorized category of users. The system integrator should determine the amount of sensitive and personal data that may be made available in the cloud that is accessible to users through the internet.

Physical and virtual controls for access need consideration in terms of enabling security mechanisms and the practice of logging instances of special access and remote access are recommended for the overall upkeep of the system. Network security centers designed to track security incidents including attempts of unauthorized users is another level of ensuring security safe operation. Simulated attacks and periodic penetration tests are part of best industry practices to check the resilience of the system against vulnerabilities.

2.4 Sources of Vulnerability

The primary source of vulnerability, via whom the entire system, or parts of it (if careful isolation methodologies are designed and implemented), is the beneficiary (patient). The plethora of devices found in the market, with associated applications for measurement of physical and biological features of the functioning of a human body, poses the real challenge to administering security while transferring data from devices to the next layer.

Patients tend to share their medical data with friends and third-party medical professionals to seek their opinion (to cross-check on their primary advice provided by the physician). Such actions expand the scope of vulnerability when login details are shared or access is made possible by proxy means to medical records.

The most common source of vulnerability is the internet itself, it being a ubiquitous platform. Hackers and attackers thrive on

1. the open access policy of the internet, and
2. on the unbridled aspirations of the development community,

whom try and experiment with new products, algorithms, and protocols. All connectors and gateways in the network are potential sources of vulnerability, and system integrations, done piecemeal, are a sure recipe to invite cyber-attacks.

Allowing access to remote users is another source of vulnerability. It is observed that when the operations and maintenance of IoHT as a service is outsourced, the vendor would like to have remote access to the system, to diagnose and rectify any alarms or malfunctioning. Granting remote access (usually through the internet) increases the risk of compromise unless the user devices are encrypted and dedicated to IoHT servicing.

Yet another method to assess and locate vulnerabilities is to deploy special techniques like forensic investigations and risk management in wireless communications [4]. Such tools help the system integrator, responsible for network security, to identify the potential points of vulnerabilities in terms of points of entry of intrusions, which would enable the designer to activate denial of service upon recognition of attempted intrusions.

2.5 Security Features

To ensure the secure and safe functioning of the IoHT as an optimal system, the following best industry practices need to be in place.

Identifying the use of the system and determining its role is a critical element among the security features. Authentication enables establishing identity and traceability, and authorization provides the scope of functions permitted to be performed by a user within the system [1].

The security policy of the institution shall designate the types of users, their roles, and responsibilities, and means of interaction. The policy should be periodically reviewed to make changes to the scope of interaction, the addition of new classes of users, and changes to the process of authentication and authorization.

Authentication enables an IoHT device to ensure the identity of the peer with which it is trying to establish access and connection. The fundamental principle to be followed is 'know your peer' by which no two nodes can establish a communication route unless the other party is identified as a transmitter or a receiver of data. In case an intermediary is designed in the network, those entities also need to be authenticated with respective labelling of the functionality they are designed to perform.

There is an emerging trend of assigning hierarchical access rights by the community of users based on the role and identity of the devices they operate by defining an operating procedure [5]. This system of authentication does not involve the central intelligence system and thus uses the community knowledge of identifying authentic users and their access rights. Such practices and protocols are recommended as part of the Industry 4.0 standards. Standards emanating from Industry 4.0 are expected to provide solutions to security issues at the edge, or nearer the action itself, thereby isolating the remote central databases and routers from possible attacks or compromises. Industry 4.0 standards along with emerging trends in cloud computing would enhance the security features of IoHT, as long as aspects of isolation and ring-fencing are considered at the design stage.

The authorization ensures that only eligible and designated nodes are accessible for network services or resources. While authentication can take place between two nodes (peers), authorization is a function to be performed by a central server, which has the global view of the network with its entire functionalities and assigned nodes. Even authenticated nodes may not be authorized to carry out functions that they have not received authorization to perform. This is significant since access to sensitive data and attempts to rewrite data could be thwarted by checking the authorization status of the device or the transaction.

The test of the resilience of the system would become evident only when interconnected health devices are compromised, yet a security mechanism should protect the network/device/information from any attack [6]. Features of isolating the parts of the network and always protecting the core (central server and storage) is a key feature of any security mechanism.

Classification of data into categories like sensitive, personal and non-personal would greatly help in designating access to various categories of users. Other than medical professionals, it suffices other users to see and transact with the kernel without having an insight into the core. An insurance company would be happy to deal with a patient identification number without seeing the personal details while processing a claim. At the time of admitting a claim, details of the patient may be verified once, rather than every time the insurance company deals with the stakeholders on the claim.

Defining a hierarchy of users, and designating the scope of access to data for each group, would help in securing the privacy of the data.

2.6 Challenges to the Security Fabric

Any common user having one device (smartphone for example) engages with the device for conducting her normal communication needs either aware of or not aware of the potential security risks associated with accessing injurious sources. The person uses the same device for IoHT, thus exposing the network/system of IoHT to the same risk as before, though the type of use and impact are variedly different. It is similar to a situation of a part of the body getting infected and the effect of infection growing and spreading to the rest of the system or interconnected network.

Providing secure access based on the principle of end-to-end is possible only when every node in the system is subject to screening and authentication in terms of its status and health (infected or not) and such a screening methodology poses a challenge when multi-modal operations of common devices are prevalent as is seen in the IoHT scenario. Protocols to scan and screen users and devices every time access is sought is both times consuming and expensive. Deployment of code of behavior and restraints of non-exposure to suspect providers form part of security mechanisms to protect against compromises. Attackers are trained to locate and identify the weakest point in the IoHT system to breach the entire network [7]. Due to the sheer volume of multi-modal entry points into the security fabric, (such as mobile devices, employee programs, cloud storage of metadata), it becomes a challenge to provide secure access to each point at the organizational level.

It is unfortunate that at times, request for change, version updates and incorporation of security fixes are not paid the necessary attention that they rightfully demand. Labeling elements and sub-systems periodically in respect of their status of update and compatibility with the rest of the components and their links, are measures to improve the security level of the system. Devices not registered with the organizations IT could be vulnerable to attacks due to outdated OS versions with no record of security patches.

Having seen that both virtual and physical breaches of protocol entail security incidents, it is of utmost importance to lend equal credence to challenges of physical access restrictions similar to what the machines do virtually. Consequential damage by either of the channels is severe and hence mechanisms to protect them should be identical and foolproof.

2.7 Security Techniques—User Authentication

Fundamental features of a robust and secure system verify and validate the user at all times. The authentication could be based on either a single step or multiple steps, depending on the application or security objective. Time-out and re-authentication measures are also recommended when the communication links are exposed to vulnerabilities. Coded and unique protocols are effective in user authentication and their deployment is to be assessed based on vulnerability tests.

ISO/IEC 9798 enlists a range of authentication protocols for different security systems and security goals [8]. System integrators and architects should evaluate the relevance of the adoption of these protocols in their system based on the perceived and assessed degree of vulnerabilities and desired resilience.

Attackers employ techniques such as camouflaging, masking and creating dummy user accounts to gain access to healthcare systems, and can hack data in the absence of security mechanisms—viz., authentication. Such deployment of various techniques by attackers demands appropriate development of complex protection systems, to counter and protect against them, on a case-to-case basis. The standards like the one mentioned above address such scenarios and effectively wards against multiple attacks.

The involvement of trusted third parties is a common practice to secure the system safety in which those parties are entrusted with the task of implementing protection systems in conformance with the mandated security standards. Trusted third parties are also required to submit the performance indices periodically to report on the types of incidents and the extent of protection afforded. The data collected by these means are used to assess the adequacy of protection including the type of standards selected for adoption. The feedback loop provided by third parties become quite useful in system upgradation and enhancement while expanding the scope of the network or increasing the number of users substantially.

ISO/IEC 9798-2:2019 provides information on entity authentication by adopting mechanisms using authenticated encryption and ISO/IEC 9798-3:2019 provides information on entity authentication adopting mechanisms using digital signature techniques [9].

The list of standards and the details thereof may scare away the system integrators and hospital administrators to consider and implement those. This is where the services of cybersecurity specialists with expertise, knowledge and skills are required to advise on the scope and extent of security measures required to be adopted and implemented. The key consideration is continuous scanning of the developments and performing cost–benefit analysis for inclusion and updating.

2.8 Conclusions

One size fits all paradigm is anathema in attempting to have a universal recommendation for establishing and regaining security and privacy of data in IoHT. The following is a set of recommendations for enabling a reasonably high level of security in the provision of services:

1. Healthcare institutions should develop and publish their approved policy on authentication and authorization of users, involved in data handling, posting and transfer.
2. Authentication and authorization, both at the physical process level and system level, should be segregated and dealt with by separate protocols.
3. Access to sensitive data should be restrained or prohibited to remote users (layers 2 and 3 of the network access).
4. An IoT device needs to be evaluated for its security protection features, concerning security criteria, before being inducted into an already existing operational environment, such as the healthcare industry. Resorting to techniques like type approval of equipment is a common practice to validate devices/equipment.

5. Healthcare institutions should implement appropriate technical and organizational measures to manage the security risks of their public networks and services. These include but are not limited to ISO22301 certification for business continuity and ISO27001 certification for information security management.
6. Security incident reporting policy and system should preferably comply with ITIL and similar global standards.
7. Healthcare institutions should regularly conduct appropriate security assessments of all application-level, server and systems software, to identify potential threats and vulnerabilities. Gaps identified during such audits/tests should be resolved in a time-bound manner.
8. Healthcare institutions should have suitable support arrangements with their vendors to ensure that security and stability fixes are provided and installed in a timely, effective manner.
9. Security protocols deployed in the system need regular updates with the latest security patches, as older versions may expose the system to new vulnerabilities. Therefore, constant updates of security patches and incorporation of request for changes are needed for IoT health devices to ward off the latest threats in the form of viruses or machine tampering. However, designing a mechanism for the dynamic scanning and installation of the latest security patches is a challenging and formidable task. Artificial intelligence will play a major contributory role in shaping policies and procedures in terms of aiding updates.
10. Healthcare institutions should conduct periodic external and internal security audits, including penetration tests that simulate real-world attack scenarios, to identify threats and weaknesses at the network operating system and application layer levels, as well as in the institution's procedures and processes.

References

1. Manyika, J., Chui, M., Bisson, P., Woetzel, J., Dobbs, R., Bughin, J., D.A., Unlocking the potential of the Internet of Things, 105–106, McKinsey Global Institute, United States, 2015, https://www.mckinsey.com/business-

functions/mckinsey-digital/our-insights/the-internet-of-things-the-value-of-digitizing-the-physical-world.

2. Islam, S.M.R., Kwak, D., Kabir, M.H., Hossain, M., Kwak, K.S., The internet of things for healthcare: A comprehensive survey. *IEEE Access*, 3, 678–708, 2015.
3. Sinha, A., Shrivastava, G., Kumar, P., Gupta, D., A community-based hierarchical user authentication scheme for Industry 4.0. Software: *Pract. Exper.*, 2020.
4. Sharma, K., Makino, M., Shrivastava, G., Agarwal, B. (Eds.), *Forensic Investigations and Risk Management in Mobile and Wireless Communications*, IGI Global, USA, 2019.
5. Sinha, A., Shrivastava, G., Kumar, P., Gupta, D., A community-based hierarchical user authentication scheme for Industry 4.0. *Software: Pract. Exper.*, 2020.
6. Joshitta R. S. M. and Arockiam L., *Device authentication mechanism for IoT enabled healthcare system*, 2017 International Conference on Algorithms, Methodology, Models and Applications in Emerging Technologies (ICAMMAET), 6, 1-6, EEE XPlore, New Jersey, Unites States, 2017.
7. Moosavi, S.R. *et al.*, SEA, A secure and efficient authentication and authorization architecture for IoT-based healthcare using smart gateways. *Proc. Comput. Sci.*, 52, 452–459, 2015.
8. Securing IoT requires a shift to a security fabric, CSO ONLINE, International Data Group CSO, Massachusetts, US, 2020 | CSO Online link, https://www.csoonline.com/article/3543073/securing-iot-requires-a-shift-to-asecurity-fabric.html.
9. ISO—ISO/IEC 9798-2:2019—IT Security techniques—Entity authentication—Part 2: Mechanisms using authenticated encryption, 4, 1–15, International Organization for Standardization, Geneva Switzerland, 2019, https://www.iso.org/standard/67114.html.

3

Security and Privacy Issues Related to Big Data-Based Ubiquitous Healthcare Systems

Jaspreet Singh

The Northcap University, Gurgaon, India

Abstract

In clinical, biomedical, and healthcare services, to continuously expand the coordination of disparate and quite different information production capabilities and to increase the availability of information in key areas of usefulness, clinics need to connect some form of manufacturer and health insurance organization so that the clinic can do it. It essentially creates a healthcare association. This data has been presented to improve well-being, increase value, and reduce costs, but the security concerns are so serious that the healthcare industry cannot use it with existing assets. However, managing and removing the logical strength of big data is an essential part of the work of any healthcare community. In this context, the purpose of this chapter is to present data security and data protection issues for big data in the healthcare industry and to talk about information security, customer access to components and methods.

***Keywords*:** Big data technology, security, privacy, HIPAA Act, privacy-preserving methods, big data in healthcare

3.1 Introduction

Essentially, Big Data has changed the way society tracks, analyzes and influences information in each industry. One of the most optimistic areas where big data is often used is Healthcare. Key health tasks can improve lasting outcomes, predict infectious diseases, add value to

Email: jsjaspreetsingh759@gmail.com

Kavita Sharma, Yogita Gigras, Vishnu Sharma, D. Jude Hemanth and Ramesh Chandra Poonia (eds.) Internet of Healthcare Things: Machine Learning for Security and Privacy, (41–64) © 2022 Scrivener Publishing LLC

experiences, prevent preventable infections, reduce medical transportation costs, and improve the quality of life for everyone. However, it can be difficult to choose the right data usage while maintaining patient safety and integrity. As useful as advances in life sciences, big data is only available when safety and integrity considerations are addressed and critical to the success of any healthcare institution. To provide a safe and stable big data environment, it is important to understand the limitations of existing solutions and lead future research. The digitization of health and patient information has undergone an emotional and central shift over a long period in clinics, work plans and actions, and primarily in finance. Population ageing and lifestyle changes the growth of mobile applications and programming; Original drugs; More attention to the quality and cost of treatment; Evidence-based drugs that are not emotional clinical options are all stimulating and provide a huge door to support clinical choice, improve treatment delivery, and provide decision-makers. Decision-makers, disease monitoring, finding harmful functions and streamlining infection treatment. It affects other body parts [1, 2]. As mentioned above, large-scale data analysis in healthcare not only offers benefits, guarantees, and tremendous potential for healthcare change but also presents complex challenges and barriers. Of course, concerns about the security of large-scale information and healthcare protection are increasing every year. In addition, medical associations have found that a flexible and innovative approach to critical security requirements is not sufficient to ensure organizational and patient safety [3].

3.2 Big Data Privacy & Security Issues

The Security & Privacy of Big Data is the main issue. Privacy is the ability to protect sensitive information from truly identifiable medical information. The focus is on the utilization and management of their information, such as developing policies to collect, share, and use patient's data, and creating approval conditions. Security is generally defined as the protection against unauthorized access, but some include certain privacy and availability notifications. It focuses on obtaining information and protecting it from malicious attacks. Security is necessary to ensure information, but providing security alone is not enough. Table 3.1 emphasizes the additional contrast between security & privacy.

Table 3.1 Difference between security & privacy.

Security	Privacy
Security is the "confidentiality, integrity & availability" of data.	Privacy is the appropriate use of user's information.
Various techniques like Encryption, Firewall, etc. are used in order to prevent data compromise from technology or vulnerabilities in the network of an organization.	The organization can't sell its patient/ user's information to a third party without prior consent of the user.
It may provide for confidentiality or protect an enterprise or agency.	It concerns with patient's right to safeguard their information from any other parties.
Security offers the ability to be confident that decisions are respected.	Privacy is the ability to decide what information an individual goes to and where to.

3.3 Big Data Security Problem

Medical facilities store, maintain, and share a wealth of information to represent productive and legitimate issues, but have the disadvantage of lack of technical support & limited security. The situation is getting more and more complicated in the medical field. Malicious hackers can use these data mining strategies and techniques to detect and discover sensitive information that could lead to data breaches. While security measures remain the avenue of communication in question, attackers are still the most vulnerable. Prices will steadily rise as methods of overcoming security measures are developed.

In terms of security and accuracy, researchers [4] argue that a high level of data security includes three aspects: information security, access control and data security. As a result, healthcare organizations must implement security controls and methods to manage appropriate security information, equipment and programming, and reliable clinical data on internal and external risks. When starting a business, you need to create a data lifecycle to make appropriate decisions about the maintenance, profitability, reuse, and analysis of registered or new data.

3.3.1 Big Data Security Lifecycle

Businesses need to implement a security policy for healthcare information that protects critical resources while complying with the conditions of reference for healthcare consistency. This lifecycle model is designed to cover the stages of the big data lifecycle and to identify threats and attacks affecting the big data domain. This stage [5] represents the big data lifecycle from the perspective of customers such as data providers, data collectors, data developers and decision makers. The proposed model [6] consisted of four connection stages: data collection stage, data storage stage, data modeling and analysis, and information generation. Researchers propose a model incorporating the measures described above to provide mechanisms and systems to combat threats and attacks during the lifecycle of Big Data. Figure 3.1 shows the major components of the Big Data Lifecycle.

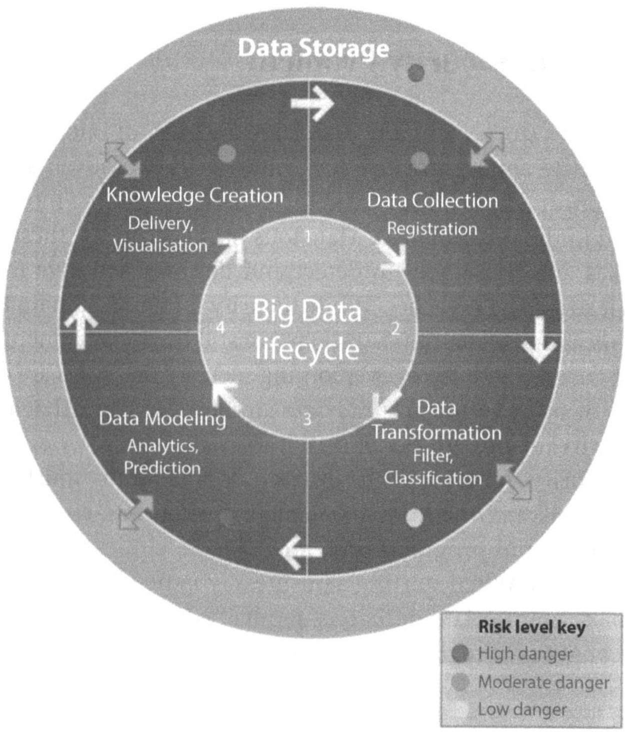

Figure 3.1 Big data security life cycle [6].

Data Collection Phase: Collect data from other sources in different organizations. From a security point of view, it is important to ensure that big data information technology becomes a critical requirement from the first stage of its life cycle. Therefore, it is important to collect reliable information. Ensure patient integrity and continuous safety. You must take carefully thought-out security measures to ensure that all information & data frames are protected from unauthorized access, duplication, modification, disclosure, redirection, misuse, destruction, loss & theft.

Data Transformation Stage: Once the data becomes available, the first step is to filter and group the data according to its structure and make significant changes to make a meaningful analysis. Data quality is also expected to improve through data filtering, enhancement and transformation. Extract data before the analysis or simulation phase, remove or manage exceptions, misplaced values, redundant data, etc. Differently, the information collected may contain sensitive data, so security is important for data transformation and security. In the information collected, the information is disabled and protected while maintaining access security and control (including user instructions, login applications, central policy archives, extensive directory and database listings such as passwords and client settings) [7] and functions. Some security measures such as data anonymity, separation and deletion.

Data Modeling Stage: When data is collected, modified and stored in secure storage, the information is carried out and sensitive information is analyzed. At this point, predictive modeling can be selected and implemented using guided data mining techniques such as grouping, classification, and linking. There are also several sets of training methods to improve the accuracy and reliability of the final model. On the other hand, it is important to create a safe working environment. Today's data processors use intuitive data grouping algorithms that can extract sensitive information. As a result, the mining cycle and network components must be designed and protected from data breaches and potential security breaches, and only qualified personnel are currently working on it. This period eliminates some weaknesses and reduces your risk for others.

Knowledge Creation Stage: In this modeling phase this stage provides new data and valuable information that decision makers can use. This information is considered confidential, especially in a competitive environment. Of course, the medical community treats information as confidential (e.g., patient personal information should not be disclosed).

Table 3.2 Threat models.

Phases	Threats & attacks	Description	Suggested defense
Data Collection	Phishing	These attacks are hacking data provider and collector to get an access to the data in the collection phase.	Security awareness program
	Spamming		
	Spoofing		
Data Storing	Data mining based attacks	Targeted datasets to extract knowledge	Divide datasets (vertically and horizontally) and non-central data storage framework.
	Attacks on data storage device	Stealing hard disks or make images of them	Physical security measures non-central data storage framework.
	Unauthorized data access	People access data illegally	Access control
Data Analytics	Data mining based attacks	Using data mining methods to extract sensitive knowledge.	Divide datasets (vertically and horizontally) and use access control.
	Re-identification threat	Identification threats of personal information	Core attribute encryption
	Wrong result threat	Using incorrect analysis process, which lead to incorrect results	Follow correct analysis procedures and document, audit, and review the process.
Knowledge creation	Privacy threats	Releasing the resulted knowledge (ex. Rival competitors)	Adopt encrypt the resulted knowledge and adopting access control strategy.
	Phishing and spoofing	Decision makers are targeted	Security awareness programs

3.3.2 Threats & Attacks on Big Data

Big Data technology is vulnerable to a variety of security threats and attacks. Table 3.2 ranks threats and attacks that vary by stage in the big data lifecycle. Each step has its properties and sets its own different goals. Therefore, each stage is exposed to different threats. In this sense, the data collection phase is vulnerable to certain attacks such as phishing and counterfeiting. These types of attacks are aimed at those who collect and present data for large amounts of data. At this stage, employees are provided with a security awareness program for data collection and how to follow security policies and practices.

Once information on storage devices is collected, police must be aware of threats and attacks, so hackers accessing data on storage devices can use data mining techniques to illegally extract and use confidential information. Data mining attack. To control the type of attack, police officers can share data sets vertically or horizontally to minimize the impact of these attacks and use a decentralized storage structure. Several threats have been identified in staged storage such as attacks on storage devices (e.g., hard disk theft) and unauthorized access attacks. In that sense, officials can build real security personnel and create access control protocols.

During the data modeling phase, some threat attacks can reveal sensitive information or affect data processing. Data mining attacks can find and publish sensitive data and use correlation strategies to identify personally identifiable information that affects personal data. To protect big data structures from these attacks, police officers can use several methods of protection. For example, splitting a data set into multiple sections (same or vertical) and encrypting the data for the underlying properties (high-pressure properties). Another threat right now is that the data analysis process can give inaccurate results. Therefore, it is important to follow the correct analysis process and register.

Finally, police officers should consider threats and attacks and how to protect them during the information generation phase. Information obtained through big data processing is considered confidential and does not need to be disclosed to the public and competitors in business situations. There are privacy threats and security attacks that can focus on decision-makers and net-benefit accessors in the big data process, so officials must also develop security policies and follow access control protocols. Software development to raise awareness. Prevent and minimize the impact of security threats.

3.3.3 Current Technologies in Use

Numerous technologies are in use to make sure the security & integrity of big health data. Commonly used techniques are:

1) Certification: Certification is the setting or confirmation that a statement about or about a company is valid and truthful and offers a fundamental opportunity in all societies. It provides access to your business, protects your customers' identities, and sees who your customers are.

Data collection can lead to rare problems, especially Man in the Middle Attacks (MITMs). Most cryptographic methods comprise endpoint authentication specifically to prevent MITM attacks. Transport Layer Security (TLS) [8] & Secure Socket Layer (SSL) are cryptographic methods that ensure communication over networks such as the Internet. SSL & TLS encrypt network connection segments at the end of the transport layer. Various versions of the protocol are widely used in applications such as Web Browsing, Email, Voice over IP (VoIP). Internet Fax & Text Messaging. Additional updates to hash methods such as SHA-256 [9] & Kerberos mechanisms are available. In addition, all sensitive 360° data can be verified using the Bull Eye algorithm. This calculation was used to ensure data security and to track the relationship between the original and the replicated data. Anyone with permission can also read or write basic information. The researchers also proposed a new and simpler validation model that uses a temporary pad algorithm. It eliminates the transmission of passwords between servers. As part of the healthcare system, health information provided by your service provider and user identity must be verified in each exam [10].

2) Encryption: Data encryption is an effective way to prevent unauthorized access to confidential information. This solution enables and maintains data logging throughout its lifecycle, from data centers to endpoints (including phones used by doctors, physicians and management). Encryption helps prevent hacks like packet sniffing and storage device theft.

Medical institutions or providers must ensure that cryptographic systems are appropriate and convenient for patients and health workers. The number of keys stored for each collection should also be limited.

Although many cryptographic algorithms have been commonly created and very well used (Rijndael, RSA, AES, RC4 [9, 11, 12], DES, RC6, 3DES [13], IDEA and so on), they are still a challenge. It determines the appropriate encryption algorithm as the best way to keep your storage secure.

3) Data Masking: Masking replaces confidential information components with unknown values. This is not an encryption strategy, so you cannot retrieve the first value from the hidden value. For example, techniques are used to anonymize records, hide personal identifiers, names, social security numbers, and prevent or aggregate semi-identifiers such as date of birth and zip code.

As a result, data hiding is one of the best-known ways to prevent real-time anonymity. K-anonymity, initially proposed by two researchers Sweeney and Samrathi [14, 15], protects against identity exposure but was not considered to protect against property exposure [16]. Other unknown methods include classes that add noise to the information, replace cells within segments, and replace groups of k records with duplicate k of one record. This method has a common problem related to the difficulty of recognizing multidimensional data sets [17, 18].

The main benefit of this approach is that it reduces the cost of protecting the use of large amounts of data. As protected data moves from secure sources to the platform, concealment reduces the need to apply additional security measures to the information contained on the platform.

4) Access Control: Once verified, clients can access the information system. However, your access will continue to be managed by an access control policy, generally based on the authority and authority of any trusted patient or trusted third party physician and a strong and customizable system for client authentication. It provides enhanced privilege control, so clients can only perform privileged operations such as accessing information, submitting jobs, organizing groups, etc.

Attribute-Based Access Control (ABAC) [20, 21] & Role-Based Access Control (RBAC) [19] are the most well-known models for electronic health records. ABAC and RBAC have emphasized some restrictions on their use. One in the medical system. Researchers [22] have also proposed effective unified access control to cryptogram cloud storage based on symmetric cryptographic algorithms (e.g., AES) for attribute encryption and cryptographic policies. It meets the requirements for detailed access control and security and privacy.

5) Monitor & Auditing: Security monitoring includes collecting and auditing network events to detect interference. Review means that customer documents are registered with the healthcare system as sequential applications, such as recording and correcting all data access. Security measures to measure and ensure the safety of medical systems [23].

Detection and prevention of interference in network traffic are complex. To solve this issue, a security surveillance architecture was created to inspect, IP Streams, DNS Traffic, HTTP Traffic [24]. The proposed solution involves storing and processing data in distributed sources using a data correlation scheme. At this point, three probabilistic measures were taken to determine whether the name of the key function, the package, or the data flow was malicious. Depending on the number of points obtained from this calculation, the alarm is triggered, triggered by the detection system, or interrupted by the suppression system.

3.4 Privacy of Big Data in Healthcare

In recent years, along with attacks on information systems, advanced threats continue to emerge, the main goal is to penetrate data recovered by attackers. Therefore, interventions on patient integrity are seen as a growing problem in large-scale data analysis, allowing organizations to address these various interrelated issues. Data security regulates access to information throughout the computing lifecycle, whereas data privacy says that access is based, for example, on the privacy policies and laws they have adopted. Anyone with access to personal, financial, clinical, or confidential information. Therefore, the confidentiality of medical information is an important factor to consider.

3.4.1 Data Protection Acts

More than ever, it is more important for healthcare organizations to track and protect personal data and to appeal to the growing jungle of relevant data protection laws, considering legal risks and obligations when processing personal data. When it comes to Data Protection, Different countries have different Law & Regulations. Data Protection Regulations & Laws in some countries are shown in Table 3.3 [25–28].

3.4.1.1 HIPAA Compliance

HIPAA stands for Health Insurance Portability and Accountablity Act. The primary purpose of this law is to make it easier for individuals to provide health insurance, to ensure the accuracy and security of health information data, and to help the healthcare industry manage regulatory costs. This law has explained requirements for the storage and confidentiality of patient information and has redefined the way safety, health and various organizations deal with health, safety and integrity issues. From an innovative perspective, HIPAA covers a broad range of areas including websites, pharmaceutical devices, medical imaging and electronic medical records.

Table 3.3 Data protection laws in some countries.

Country	Law	Salient features
U.S.A.	HIPAA Act Patient Safety and Quality Improvement Act (PSQIA)HITECH Act	Requires the establishment of national standards for electronic healthcare transactions. Gives the right to privacy to individuals from ages 12 through 18. Signed disclosure from the affected before giving out any information on provided healthcare to anyone, including parents. Patient Safety Work Product must not be disclosed. Individual violating the confidentiality provisions is subject to a civil penalty. Protect security and privacy of electronic health information.
EU	Data Protection Directive	Protect people's fundamental rights and freedoms and in particular their right to privacy with respect to the processing of personal data.
Canada	Personal Information Protection and Electronic Documents Act ('PIPEDA')	Individual is given the right to know the reasons for collection or use of personal information, so that organizations are required to protect this information in a reasonable and secure way
UK	Data Protection Act (DPA)	Provides a way for individuals to control information about themselves. Personal data shall not be transferred to a country or territory outside the European Economic Area unless that country or territory ensures an adequate level of protection for the rights and freedoms of data subjects.

(Continued)

Table 3.3 Data Protection Laws in some countries. (*Continued*)

Country	Law	Salient features
Morocco	The 09-08 act, dated on 18 February 2009	Protects the one's privacy through the establishment of the CNDP authority by limiting the use of personal and sensitive data using the data controllers in any data processing operation
Russia	Russian Federal Law on Personal Data	Requires data operators to take "all the necessary organizational and technical measures required for protecting personal data against unlawful or accidental access"
India	IT Act and IT (Amendment) Act	Implement reasonable security practices for sensitive personal data or information. Provides for compensation to person affected by wrongful loss or wrongful gain. Provides for imprisonment and/ or fine for a person who causes wrongful loss or wrongful gain by disclosing personal information of another person while providing services under the terms of lawful contract
Brazil	Constitution	The intimacy, private life, honor and image of the people are inviolable, with assured right to indigenization by material or moral damage resulting from its violation
Angola	Data Protection Law (Law no. 22/11 of 17 June)	With respect to sensitive data processing, collection and processing is only allowed where there is a legal provision allowing such processing and prior authorization from the APD is obtained

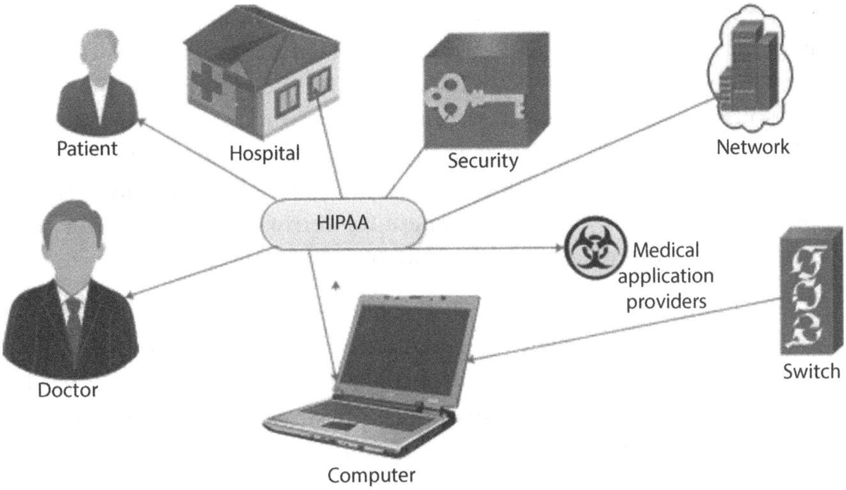

Figure 3.2 General HIPAA Diagram [47].

Various modern innovation models, including virtualization, distributed computing, mobile device adoption, and apps, have presented new challenges for HIPAA compliance for payers and caregivers (Figure 3.2).

Solutions can assist organizations to meet HIPAA requirements and improve their overall safety and risk management position. These mechanisms provide strong protection against threats from client sites, mobile phones and many more. It includes a brief overview of the impact on different areas of innovation and the impact of trends such as cloud and mobility on compliance.

3.4.1.2 HIPAA Five Rules

1) Privacy Rule: HIPPA's Privacy Policy defines who can access Protected Health Information (PHI). These rules apply to all forms of communication, including oral, written or electronic disclosure. Mainly, patients must sign an association waiver. Without written consent, protected persons may not disclose personally identifiable information to anyone who does not fall under the Permitted Use and Disclosure section of HIPAA. Companies tracked include health insurance plans, healthcare providers, and health information centers.

Protected devices can disclose PHI to anyone with whom they share data. Indoor devices may also use and disclose patient data needed to support operations, including treatment and payment procedures. Some caregivers have patient catalogues, so guests who ask for the patient's name will know where the patient is in the hospital and will provide their health status. Providers may use the patient's unintended consent to disclose information to people who want to pay for services or to make plans for the patient's death.

Confidentiality rules also allow clinical organizations to share data with public health authorities that have the legal authority to obtain information to prevent or manage illness, injury, or disability. As public health authorities deal with incidents, prepare for terrorism, and monitor programs, research on personal health data may be required.

2) Security Rule: Describes the rules, specifications, and security cycles for electronic PHI. It defines how protected health information (PHI) should be communicated and does not play a role in oral communication. Safety rules include specifications for administrative, physical and technical controls. Identify specific individuals or groups to implement and monitor compliance within the organization. To comply with HIPAA, indoor organizations must use firewalls, encryption, modern software, and operating systems

a) **Administrative safeguard:** There is a necessity to practice, create, and maintain policies. The policies and methods used to educate and support advocates to maintain PHI security must be implemented, created, and maintained. Some examples of administrative security measures include:

(i) Accept the use of the policy to help the mentor or representative gain authority and responsibility for the transfer of PHI.
(ii) A penalty policy is required to educate employees who violate HIPAA.
(iii) Access to information policy provides appropriate access to PC workstations, medical and shared registers, and various projects or cycles.
(iv) Safety training is required. In this way, employees are trained and assisted to remember policies and procedures identified through software updates, PC log checks, password updates and other important security measures.
(v) An emergency plan to ensure that appropriate policies and procedures are in place to prepare for emergency response. In the event of a fire, vandalism or other catastrophic events, you must develop, test and update your accident and emergency plans and designate all important exercises as owners.

b) **Technical safeguards:** Precautions: Exercise requires proper procedures and software and hardware to protect PHI. Practitioners should update their policies and technical methods to provide access to everyone who needs access to manage their responsibilities. Practitioners need to combine encryption and decoding and electronically back up, restore and transmit patient data. PHI destruction policies and methods should be developed when a job or function is no longer needed.

c) **Physical safeguards:** It must be followed to ensure that the zones and units are close to your work. Access to facilities must be controlled and all access must be respected. It is important to understand and track the people involved in the training and to take safety measures before and after potential accidents. HIPAA requires a HIPAA Security Officer and Privacy Officer to be appointed at each training session. Security officers are trained in various organizations, as shown in Figure 3.3.

3) **Transaction Rule:** A transaction is an electronic exchange in which data is transferred between two units for a specific reason, and the medical office

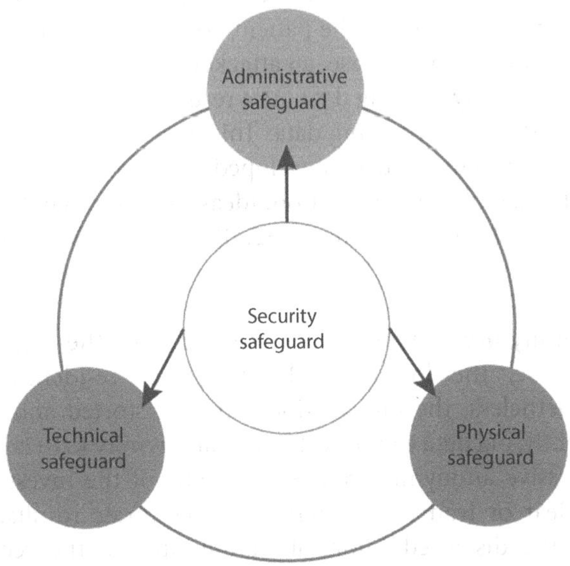

Figure 3.3 Process of HIPAA.

sends the matter to the health insurance plan through the clearinghouse to require payment for medical services provided. Operated by HIPAA. List of allowed transactions and rules for transmitting PHI. If your insurance company conducts one of these transactions electronically, you must follow HIPAA rules. Diagnosis and methods.

4) Unique Identifier Rule: To simplify HIPAA management, three identifiers are used to represent indoor objects throughout the transaction. This allows you to standardize processes, reduce errors and increase efficiency.
5) Enforcement Rule: This rule encourages caregivers to update electronic medical or EHR records. Incentives and cash aids are provided through the HITECH Act, allowing caregivers to improve the effectiveness of their services.

3.5 Privacy Conserving Methods in Big Data

Some of the practices of data privacy are described here. These procedures are often used to ensure patient confidentiality [29–31], but their drawbacks have led to new approaches.

a) Anonymization: Anonymization is a method of limiting the disclosure of personal data by discarding potentially disruptive data either through primary methods requiring removal of certain patient identifiers or through always measurable strategies. The patient-controlled himself enough. The identifier is removed. In any case, an attacker could get additional help on external data to anonymize big data. As a result, anonymization alone cannot guarantee the security of big data. This is more reliable if an effective data protection system has been developed to maintain an algorithm that minimizes the risk of re-identification. Ideas for k-anonymity [15, 32, 33], l-diversity [33–35] and t-proximity [32, 35] were introduced to improve this technique.

- **k-anonymity:** When using this method, the higher the k value, the lower the likelihood of re-identification. Nonetheless, the information can be distorted and consequently, big data can be lost due to anonymity k. In addition, excessive anonymity can render some of the investigation unclear or lead to unilateral and inaccurate results, making the disclosed information unusable to the recipient. Sensitive properties (e.g. disease) are now revealed as one

Table 3.4 An anonymous database containing patient records.

Name	Birth	Sex	ZIP code	Religion	Disease
John	01/08/1988	Male	20056	Christian	Viral Infection
Ricky	30/01/1964	Male	20303	Christian	Cancer
Khalid	21/01/1972	Male	20043	Muslim	Heart-related
Yasmine	12/03/1969	Female	20152	Muslim	Diabetes
Aisha	15/09/1982	Female	20042	Hindu	Dengue

of the following when using semiconductors that contain information in k-address to identify a person by interacting with other publicly available information: Identifier. Table 3.4 is an anonymous database containing patient records from an emergency simulation clinic in Camelot.

This data has six columns and five items. There are two usual ways to get the anonymity of k for k values.

The first way is suppression, in this way the reference cursor '*' can override some attribute values. You can replace all or part of the meaning of the section with "*". In Table 3.5 anonymous, all attributes of the "Name" and all values of "Religion" are specified in "*" are replaced.

The second approach is to generalize, in this strategy, the values for individual features are replaced by broader categories. The "Birth" attribute can be replaced with the "1969" year. The zip code field is summarized to cover a larger display area (Camelot).

Table 3.5 Two anonymities with properties "birth", "sex" and "zip".

Name	Birth	Sex	ZIP code	Religion	Disease
*	1988	Male	20000	*	Viral Infection
*	1954	Male	20000	*	Cancer
*	1972	Male	20000	*	Heart-related
*	1969	Female	20000	*	Diabetes
*	1982	Female	20000	*	Dengue

Table 3.5 has two anonymities for the "birth", "sex" and "zip" attributes. This is because, for each combination of these attributes, there will always be more than one of these attributes on exactly one row. At the table, the double "half-identifier" still occurs in k-data records to register data with k-anonymity. Anonymized K information can in any case be immune to attacks such as unaligned pair attacks, time attacks, and additional drop attacks [14, 35]. The odd thing is that there is a difficulty in mapping the relationship between personal records and anonymity in terms of limiting the amount of undisclosed data, ensuring personal anonymity with k collection size, and maintaining database lines to achieve this security. This optimization problem is NP-Hard [36].

Numerous measures have been suggested to assess data loss due to anonymity, but do not reflect the actual usefulness of the information [37, 38]. Therefore, researchers use strategies to anonymize L-diversity data.

- **L-diversity:** A form of anonymity within a group used to protect the confidentiality of data collection by reducing the granularity of the data presented. This model (Notable, Entropy, Recycling) [14, 32, 33] is an extension of k-anonymity that uses strategies involving generalization and inhibition to reduce the roughness of data. The L-diversity model eliminates some drawbacks of k-anonymity. One model that guarantees an individual is a k-individual to the extent that does not constitute the protection of the associated secret values that have been accumulated or repressed. The problem with this strategy is that it relies on a range of sensitive properties. The information should be L-fold, but the sensitive attribute doesn't mean as much as the fake data should contain. While this false information increases security, it can cause problems at the center of the analysis. Therefore, asymmetric and pseudo-attack [35] is responsible for L-diversity and cannot be prevented from revealing attributes.
- **T-closeness:** It is an Advanced model for transformation based on group diversity. The proximity model t (correspondence/progress section) [32, 35] extends the diversity model l by processing specific attribute values concerning the circulation of information values for that attribute. The biggest improvement of this method is that it mediates attribute expansion, and as the size and diversity of the data increases, so does the likelihood of re-identification.

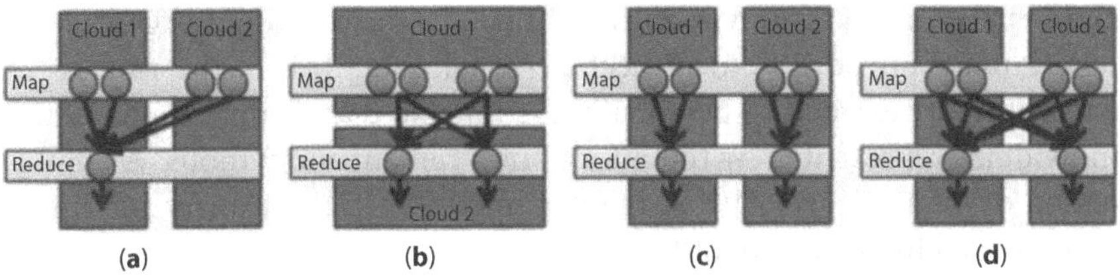

Figure 3.4 The four Operating Execution categories for HybrEx MapReduce [40]. a) Map hybrid; b) Horizontal Partitioning; c) Vertical Partitioning; d) Hybrid.

b) HybrEx: The Hybrid Execution Model [39] is an integrity and security model for cloud services. It uses the public cloud only for the organization's unclassified information and computers. When exchanging information and performing calculations using the public cloud, and for the organization's private and confidential information and calculations, the model runs a private cloud. In addition, when the application expects access to public and private information, the application itself is also shared and runs in public and private clouds. Consider the integrity of your information and ensure secure integration before completing the assignment.

Four categories of HybrEx MapReduce with new types of applications using public and private clouds are shown in Figure 3.4.

1. Map Hybrid: Map steps are performed in public & private clouds, and reduced steps are performed only in one of the clouds.
2. Vertical Partitioning: Matching & Truncation operations are performed in the public cloud using public operations such as entering and storing results in the public cloud. A similar operation is performed on a private cloud with personal information. The work is prepared by yourself.
3. Horizontal Partitioning: This map phase is executed in the public cloud & the downgrade runs on a private cloud.
4. Hybrid Stage: Map and downgrade phases work in both public and private clouds. Data transfer between clouds is also possible.

The problem with HybrEx is that it controls the cloud as an enemy rather than controlling the keys that appear in the public and private clouds at the map stage.

c) Identity Based Anonymization: This is a form of data organization whose main purpose is to protect personal data. This is the process of encrypting or extracting personally identifiable data from a dataset, and individuals represented by the information remain anonymous. A combination of naming, privacy and big data methods [41] for analyzing information used to protect identity.

For example, a group of Intel Human Factor Engineers wanted to protect Internet representatives by improving the usability of the actively used Intel in-house web portal using data logs and large websites. Personal Identifiable Information (PII) had to be deleted. Archive the portal deployment log in a way that does not interfere with the use of big data analytics tools or the ability to re-enter log entries to investigate unusual behavior.

To take advantage of the tremendous benefits of Cloud Storage [42], Intel created an open architecture for anonymity [41], allowing weblog posts to be anonymized and identified using a variety of tools. Company data has its characteristics compared to the standard practice of anonymity [43]. Intel also concealed visible personal information such as usernames and IP addresses but found that anonymous information was not protected from correlation attacks. By investigating the pros and cons of these vulnerability tests, we found that there is a strong connection between user agent data and individual users. It is a company's anonymity use case study that explains the need, behavior, and experience of the meeting when using anonymity to protect the integrity of big business data analyzed using a big data strategy. k-Anonymity based on anonymity d. Intel has also used Hadoop to explore anonymous information and generate meaningful results for human factor analysts [44, 45]. Simultaneously, they discovered that anonymity required more than hiding or summarizing certain areas. Anonymous data sets must be carefully analyzed to determine their vulnerability to attack.

3.6 Conclusion

The potential of big data in healthcare is endless, but there are some obstacles to its true potential, including technical difficulties. Big Data Security and Data Protection are considered major obstacles for analysts in this field.

This chapter has outlined the privacy and security problems at each stage of the big data lifecycle, as well as the advantages and disadvantages that arise in the current evolution of Big Health Data Security & Privacy.

We discussed the various laws in different countries. We also discussed the five rules of HIPAA.

We mainly looked at the latest confidentiality practices in healthcare and discussed how to use encryption and anonymity strategies to ensure accuracy and limitations in healthcare. There are also a variety of methods, including Haystack Needle Protection [46], Access Control, Homogeneous Encryption, & Storage Path Encryption. Nevertheless, problems always arise.

In this scenario, the point of view is to find a profitable solution for data protection and security under key health information, as in the future process. We also need to improve our data protection practices.

With the advancement of the Internet of Things, the clearer the quantity, the lower the quality. As a result, the quality of information does not have to rely on privacy algorithms to get an appropriate response from the analyst. Researchers are trying to solve the problem of adapting to the security and integrity model by encourages a variety of ways that guide a decision-making and planning approach to final support work.

References

1. Burghard, C., *Big data and analytics key to accountable care success*, IDC Health Insights, Framingham, 2012.
2. Fernandes, L., O'Connor, M., Weaver, V., Big data, bigger outcomes. *J. AHIMA*, 83, 38–42, 2012.
3. David Houlding, MSc, CISSP. *Health Information at Risk: Successful Strategies for Healthcare Security and Privacy*, Healthcare IT Program of Intel Corporation, white paper, 2011.
4. Kim, S.-H., Kim, N.-U., Chung, T.-M., Attribute relationship evaluation methodology for big data security, in: *2013 International Conference on IT Convergence and Security (ICITCS)*, IEEE, pp. 1–4, https://doi.org/10.1109/icitcs.2013.6717808.
5. Xu, L., Jiang, C., Wang, J., Yuan, J., Ren, Y., Information security in big data: Privacy and data mining. *J. Rapid Open Access Publ.*, 2, 1149–76, 2014, https://doi.org/10.1109/ACCESS.2014.2362522.
6. Yazan, A., Yong, W., Raj, K.N., Big data life cycle: threats and security model, in: *21st Americas Conference on Information Systems*, 2015.
7. General Dynamics Health Solutions white paper UK. Securing Big Health Data ©2015. http://gdhealth.com/globalassets/healthsolutions/documents/brochures/securing-big-health-data_-white-paper_UK.pdf, 2015.

8. Zhang, R. and Liu, L., Security models and requirements for healthcare application clouds, in: *IEEE 3rd International Conference on Cloud Computing*, 2010.
9. Shafer, J., Rixner, S., Cox, A.L., The Hadoop distributed filesystem: Balancing portability and performance, in: *Proceedings of 2010 IEEE International Symposium on Performance Analysis of Systems & Software (ISPASS)*, White Plain, NY, pp. 122–33, March 2010.
10. Yang, C., Lin, W., Liu, M., A novel triple encryption scheme for hadoop-based cloud data security, in: *Emerging Intelligent Data and Web Technologies (EIDWT), 2013 Fourth International Conference on*, pp. 437–42, 2013.
11. Federal Information Processing Standards Publication 197. Specification for the advanced encryption standards (AES), November 2001.
12. Somu, N., Gangaa, A., Sriram, V.S., Authentication service in hadoop using one-time pad. *Indian J. Sci. Technol.*, 7, 56–62, 2014.
13. Fluhrer, S., Mantin, I., Shamir, A., Weakness in the key scheduling algorithm of RC4, in: *8th Annual International Workshop on Selected Areas in Cryptography*, Springer-Verlag, London, 2001.
14. Sweeney, L., Achieving k-anonymity privacy protection using generalization and suppression. *Int. J. Uncertain Fuzz. Knowl. Based Syst.*, 10, 571–88, 2002.
15. Samrati, P., Protecting respondents' identities in microdata release. *IEEE Trans. Knowl. Data Eng.*, 13, 1010–27, 2001.
16. Truta, T.M. and Vinay, B., Privacy protection: p-sensitive k-anonymity property, in: *Proceedings of 22nd International Conference on Data Engineering Workshops*, p. 94, 2006.
17. Spruill, N., The confidentiality and analytic usefulness of masked business microdata. *Proceedings on Survey Research Methods*, pp. 602–607, 1983.
18. Chawala, S., Dwork, C., Sheny, F.M., Smith, A., Wee, H., Towards privacy in public databases, in: *Proceedings on Second Theory of Cryptography Conference*, 2005.
19. Science Applications International Corporation (SAIC). Role-based access control (RBAC) Role Engineering Process, Version 3.0, P. 4–7, 2004.
20. Mohan, A., Blough, D.M., An attribute-based authorization policy framework with dynamic conflict resolution, in: *Proceedings of the 9th Symposium on Identity and Trust on the Internet*, 2010.
21. Hagner, M., Security infrastructure and national patent summary, in: *Tromso Telemedicine and eHealth Conference*, 2007.
22. Zhou, H., Wen, Q., Data security accessing for HDFS based on attribute-group in cloud computing, in: *International Conference on Logistics Engineering, Management and Computer Science (LEMCS 2014)*, 2014.
23. Linden, H., Kalra, D., Hasman, A., Talmon, J., Inter-organization future proof HER systems—A review of the security and privacy related issues. *Int. J. Med. Inform.*, 78, 141–60, 2009.

24. Marchal, S., Xiuyan, J., State, R., Engel, T., A big data architecture for large scale security monitoring. *Big Data (BigData Congress)*, Anchorage, AK, pp. 56–63, 2014.
25. Piper, D. L. A. Data protection laws of the world: full handbook. DLA Piper 1, 1–50, 2017.
26. Magzoub, A., Privacy and Big Data, by Craig, T., and Ludloff, M. E., *Information Security Journal: A Global Perspective*, 21, 6, 353–354, 2012.
27. Jensen, M., Challenges of privacy protection in big data analytics. *2013 IEEE International Congress on Big Data*, 2013.
28. Chafiol-Chaumont, F. and Falkman, A.-L., *Data protection overview (Morocco)*, 2013.
29. Patil, H.K. and Seshadri, R., *Big Data Security and Privacy Issues in Healthcare*. In Proceedings of the 2014 IEEE International Congress on Big Data (BIGDATACONGRESS '14). 762–765, IEEE Computer Society, USA, 2014. https://doi.org/10.1109/BigData.Congress.2014.112
30. Sectorial Healthcare Strategy 2012–2016. Moroccan Healthcare Ministry.
31. Patil, P., Raul, R., Maurya, R.S.M., *Big Data in Healthcare*, International Journal of Research in Information Technology, 2, 2, 202–208, February 2014.
32. Li, N. *et al.*, t-Closeness: privacy beyond k-anonymity and L-diversity, in: *Data Engineering (ICDE) IEEE 23rd International Conference*, 2007.
33. Ton, A., Saravanan, M, Ericsson research, http://www.ericsson.com/research-blog/data-knowledge/big-data-privacy-preservation/2015.
34. Machanavajjhala, A., Gehrke, J., Kifer, D., Venkitasubramaniam, M., L-diversity: Privacy beyond k-anonymity, in: *Proc. 22nd International Conference Data Engineering (ICDE)*, p. 24, 2006.
35. Samarati, P. and Sweeney, L., Protecting privacy when disclosing information: k-anonymity and its enforcement through generalization and suppression, Technical Report SRI-CSL 98-04, SRI Computer Science Laboratory, 1998.
36. Meyerson, A. and Williams, R., On the complexity of optimal k-anonymity, in: *Proc. of the ACM Symp. on Principles of Database Systems*, 2004.
37. Iyenger, V., Transforming data to satisfy privacy constraints, in: *Proceedings of the ACM SIGKDD*, pp. 279–88, 2002.
38. LeFevre, K., Ramakrishnan, R., DeWitt, D.J., Mondrian multidimensional k-anonymity, in: *Proceedings of the ICDE*, p. 25, 2006.
39. Priyank, J., Manasi, G., Nilay, K., Big data privacy: a technological perspective and review. *J. Big Data*, 3, 25, 2016.
40. Ko, S.Y., Jeon, K., Morales, R., The HybrEx model for confidentiality and privacy in cloud computing, in: *3rd USENIX Workshop on Hot Topics in Cloud Computing, HotCloud'11*, Portland, 2011.
41. Sedayao, J. and Bhardwaj, R., Making big data, privacy, and anonymization work together in the enterprise: Experiences and issues, in: *Big Data Congress*, 2014.

42. Yong, Y. *et al.*, Cloud data integrity checking with an identity-based auditing mechanism from RSA. *Future Gen. Comput. Syst.*, 62, 85–91, 2016.
43. Dijcks, J.-P., Big Data for the Enterprise, Oracle, October 2011.
44. Hadoop Tutorials, 2012, https://developer.yahoo.com/hadoop/tutorial.
45. Hadoop Fair Scheduler Design Document (PDF). apache.org. Retrieved 12 October 2017.
46. Jung, K., Park, S., Park, S., *Hiding a needle in a haystack: privacy preserving Apriori algorithm in MapReduce framework PSBD'14*, Shanghai, pp. 11–7, 2014.
47. Mbonihankuye, S., Nkunzimana, A., Ndagijimana, A., Healthcare Data Security Technology: HIPAA Compliance, Wireless Communications and Mobile Computing, vol. 2019, Article ID 1927495, 7 pages, 2019. https://doi.org/10.1155/2019/1927495

Section 2

APPLICATION OF MACHINE LEARNING, BLOCKCHAIN AND FOG COMPUTING ON IoHT

4

Machine Learning Aspects for Trustworthy Internet of Healthcare Things

Pradeep Bedi[1], S.B. Goyal[2*], Jugnesh Kumar[3] and Preetishree Patnaik[3]

[1]Lingayas Vidyapeeth, Faridabad, India
[2]City University, Petaling Jaya, Malaysia
[3]St. Andrews Institute of Technology & Management, Gurgaon, India

Abstract
With the rapid development of communication tools and techniques as well as intelligent processing and analysis had led to the development of smart Internet of Things (IoT) applications. The IoT application generates a huge amount of data every day as its application is vast. One of the most important applications of IoT is in the healthcare sector that generates large and sensitive data. So, to process such a large amount of data over an insecure network is an issue of concern. In this chapter, an overview of IoT architecture, issues, and solutions are discussed. The existing privacy-preserving inference approaches have problems such as high computation and communication overheads. In this chapter an overview of security challenges and security models using machine learning for the Internet of Healthcare Things (IoHT) is given along with a framework is proposed that can enforce security and trustworthiness on the Internet of Healthcare Things (IoHT) by amalgamating privacy and deep learning. This chapter will give a direction for future research work while designing a secure IoHT framework with low latency, and fast processing for accurate end-to-end data delivery.

Keywords: Internet of Things, healthcare, security, trust, intrusion detection, machine learning

**Corresponding author*: drsbgoyal@gmail.com

Kavita Sharma, Yogita Gigras, Vishnu Sharma, D. Jude Hemanth and Ramesh Chandra Poonia (eds.) Internet of Healthcare Things: Machine Learning for Security and Privacy, (67–94) © 2022 Scrivener Publishing LLC

4.1 Introduction

With increasing application in our daily lives, the Internet of things (IoT) had become one of the most contributing applications. Increasing IoT applications has accelerated our movement from the physical world to the virtual world. Different heterogeneous things or sensors communicated with each other with the integration of IoT. Out of several applications of IoT, the healthcare sector is one of the most promising applications.

On the Internet of Healthcare Things (IoHT) applications, IoT smart sensor nodes, and interconnected networks are deployed for the health assessment of patients. Generally, patients suffering from chronic diseases such as heart disease, diabetes, hypertension, etc. need emergency services as compared to regular treatments. Internet of Things (IoT) collects data from sensing devices through middleware. To handle effectively such heterogeneous data and provide secure transmission over middleware is an area of concern and it is a quite challenging task [1–5].

While dealing with sensitive information of patient's data over an insecure channel, it is required to ensure security and privacy. During synchronization of smart sensors, gateway, or storage servers, privacy is a major concern in healthcare applications. Some of the major challenges are discussed as below [6]:

- Failure of sensing nodes randomly;
- Privacy while collecting data;
- Disruption while transmission of data;
- Malware attack;
- Intrusion.

According to recent research works, a lot of security issues have been focused to ensure cybersecurity or network security. That's why it is required to design and develop an approach that takes these security concerns into account and understand all aspects.

This chapter is dedicated to focusing light on security issues on the Internet of Healthcare Things (IoHT). In Section 4.2, an overview of IoT is discussed with their application areas. Section 4.3 discusses the issues and challenges of IoT for understanding issues of IoHT as security concern remains the same under all applications of IoT. Section 4.4 describes IoHT architecture and its challenges. Section 4.5 describes the security protocols that are existing in the current scenario. In Section 4.6, different machine learning approaches are discussed for the IoHT environment. In Section 4.7, after analyzing the drawbacks of the existing system, a secure framework for IoHT is proposed along with its key advantages. Finally, in Section 4.8, the conclusion and future research scope of this chapter are described.

4.2 Overview of Internet of Things

From the technical point of view, the Internet of Things (IoT) is simply termed as the conversion of physical devices, objects or things into a digital product. This brings a new vision or meaning to real-world things by making them more intelligent. This means adding computing capabilities to all things or devices either connected to cloud end servers or not as well as integrating back-end application support for viewing and analyzing data and controlling those devices. IoT is a term that anticipates the communication between objects in everyday life by utilizing various kinds of sensors such as Radio-Frequency Identification (RFID), actuators that function together to detect, capture and transfer essential data to the Internet from their setting. IoT is a concept that, through using happy technology, envisions communication between the physical and digital worlds. Owing to the wide size of the digital context, the number of users and the number of smartphones, energy consumption rates have reached distressing levels over the past decade. The growth in the number of linked devices by 2030 would be up to 100 billion [7]. Scientists however expect a massive information rate and an immense information size at the expense of the extraordinary release of carbon into the atmosphere. In later years, it is anticipated to grow. Clean or green energy is becoming an attractive study field in the advancement of technology because of these immense (CO_2) pollution, environmental and health issues. One of the fascinating innovations is the Internet of Things (IoT) [8]. It makes it possible to connect people and stuff everywhere, wherever, with anybody and everything, utilizing any connection and any service. It provides a forum for seamlessly connecting devices and sensors inside a smart world to provide people with highly sophisticated services. All types of data about the target are sensed and collected by sensors and devices and then the data could be further processed and analysed to retrieve relevant detail to make smart utilities [9].

In basic, there are four key components in IoT technology as described in Table 4.1:

Hardware: Integrated hardware including some sensors, tags, actuators and transceivers for information exchange. The devices in these units can share common data irrespective of their heterogeneous behaviour or environment. The data surrounding them are sensed, collected and transmitted to the next layer i.e., the network layer for further communication through gateways or routers. The devices such as sensors or actuators can communicate with each other locally through coordinating devices [10].

Table 4.1 Basic architectural layers of IoT.

Layers	Description
Hardware	This unit consists of: Components such as Sensors, RFIDs, etc.
Internet	This unit consists of: communication Interfaces.
Presentation	This unit consists of: Data storage units, Data analytical tools and applications. Platform for remote access of end-users.

Middleware: For storing data, computation, and knowledge of content intermediately.
Internet: Provides connectivity between anything at anytime, anyplace. It includes cloud computing for intelligent devices, intelligent web applications, and IP.
Presentation or Application: For various platforms and applications to grasp visualization and interpretation software.

Several areas covered by IoT applications are given in [11]:

- IoT Devices Control: In this unit, all functions and procedures are included that are needed for control for IoT devices.
- Control Operations: In this section, all operational control such as data analysis, monitoring and management are performed by IoT devices.
- Collection of Information: In this section, the collection of data, their transmission to server and finally response of decision-making units are represented.
- Application Interface: This unit represents the interface or programming application to which the user interacts.
- Business Interactive Applications: This unit merges the IoT applications and functions. For example, Customer Relationship Management (CRM) or Enterprise Resource Planning (ERP) are some examples of business applications.

In most IoT application, cloud servers are integrated as backend support either locally or remotely. These cloud servers can be physical or virtual while supporting IoT applications. An IoT platform consists of applications for operations, business, and user interface that consists of controlled data flow among them. Apart from this, the IoT platform also supports remote applications and remote access of data or resources outside the IoT environment. According to necessity, the architecture of IoT can be defined in different ways, as illustrated in Figure 4.1. Among all available standards, ISO/IEC/IEEE standards [11] are considered to be the basic architecture for IoT application deployment.

To understand the potential of IoT application, first of all, it is necessary to understand the available reference architectures of IoT stated by different organizations. According to application need, it is required to adopt or change the IoT architecture. So, it can be stated that there is no fixed architecture of IoT, as it is dependent on the application area. Therefore, many recent research works are utilizing their efforts while designing the architecture of IoT to achieve scalability, security, energy efficiency, etc.

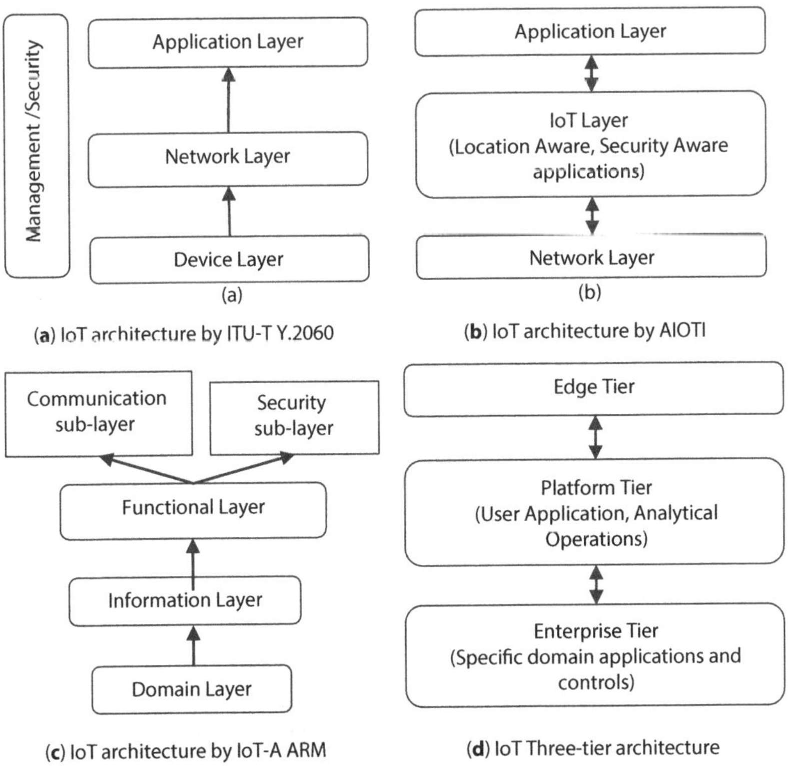

Figure 4.1 Architectures of IoT based on applications [12].

There are different designs available based on parameters such as cost, dynamic adaptability, security, energy efficiency, and latency. These factors determine the architecture of IoT, for example, no-OS, Real-time OS, full-OS, server-OS, serverless-OS, etc.

4.2.1 Application Area of IoT

IoT systems can perform remote sensing, intervention, and monitoring support. The IoT machines should be made smart enough to operate without any intervention from human beings. IoT aims to provide strong connections between real-life applications and computer applications and can contribute to improvements in economic wellbeing and accuracy and reliability and minimize human intervention. Figure 4.2 illustrates the general architecture of the IoT network. Many intelligent instruments, including sensors and actuators, are installed in various structures. All IoT intelligent devices

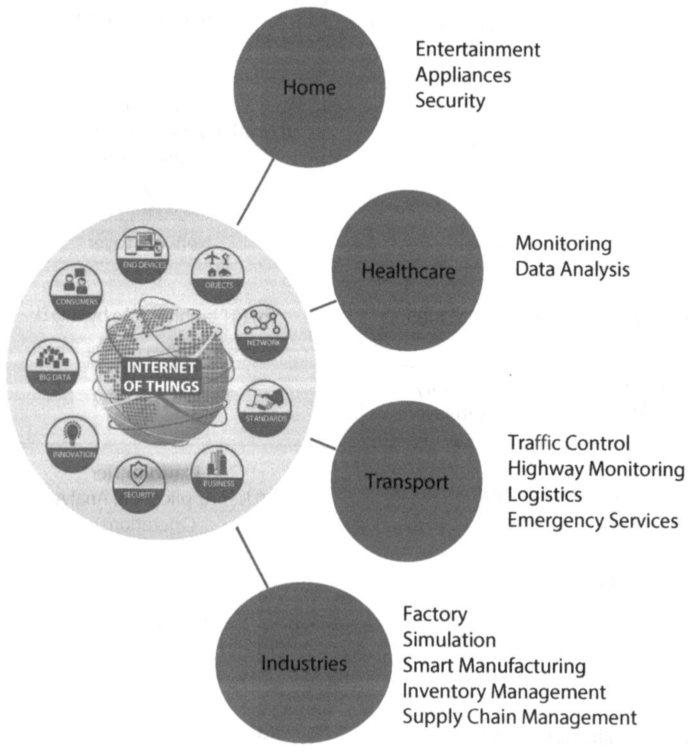

Figure 4.2 IoT applications.

are connected to the internet by trusted Gateway Nodes (GWNs). The data accessed by IoT computers can also be used by various applications.

4.2.1.1 *Wearable Devices*

Health monitoring using smart sensors is one of the flagship IoHT apps. It is designed to treat patients with diabetes. It tests a glucose level in the patient's body through a small electrode called a "glucose sensor" mounted under the skin. Wearable devices such as "Apply Bits," 'heart rate tracking' and intelligent reloaders are very popular these days and transmit the collected information to the associated monitoring system via radiofrequency. Also available are several types of devices that can be easily carried by the individual that includes Guardian glucose monitoring as one of its key example [13–16].

4.2.1.2 *Smart Home Applications*

Smart House forms a leading and evolving application of the IoT based monitoring systems in the recent arena. An intelligent home is designed for the remote control of lighting, heating, cooling, and other electrical appliances using a smartphone or computer system. The best example is "Jarvis," which is an intelligent, artificial intelligence-based home automation system (AI) [17].

4.2.1.3 *Healthcare IoT Applications*

With the application of IoT, the healthcare system is improved from reactive to proactive. Smart sensor nodes or medical appliances are deployed to capture data from patients and send it to the remote server for processing (cloud server). If any user or specialist (doctors) wants to access the information then they can access it through a cloud server remotely. This integration of IoT in the healthcare sector has improvised the efficiency level in real time [18].

4.2.1.4 *Smart Cities*

Most towns are currently attempting to transform their towns into intelligent cities in several countries. There are features of a clever city like intelligent houses, intelligent traffic management, etc. With the increase in smart features, their challenges come together. Various towns are faced with different problems (e.g., reduced clean water supply, higher population densities, and lower air quality index) which occur at different forms and paces in different areas. Consequently, these factors affect an individual area in a particular way. The agencies concerned shall then use the IoT environment to determine these complex community planning considerations by city. Usage of IoT equipment can help facilitate diverse application areas, smart monitoring, and surveillance, traffic management, etc. [19–22].

4.2.1.5 *Smart Agriculture*

The world's population will grow every day to surpass nearly 10 billion by 2050. Then it will become a challenging task to cultivate and provide food to everyone. Our cultivation practices must also be modified. New technologies can be used, such as "Smart Greenhouse" The greenhouse farming method increases crop yield by controlling the environmental factors that can influence cultivation. Although managing causes a lack of production and loss of energy, high labor costs reduce the efficiency of the whole operation. The greenhouse approach uses smart interconnected devices that promote the tracking and control of environmental variables such as humidity, pH balance, nutrients content, in the crop area [22, 23].

4.2.1.6 *Industrial Internet of Things*

Another application of IoT is the Industrial Internet of Things (IIoT) which makes our manufacturing and production system smart with the integration of smart machine and monitoring systems. As the unexpected downtimes and machine faults in such an area trigger human causality, the manual fault diagnosis and correction system is replaced with smart sensors that can predict and track the fault. These smart systems will generate the alarm that is intended to take immediate steps to avoid any casualties at the industrial site [24].

4.3 Security Issues of IoT

In this section, we discuss different security requirements (Figure 4.3) in IoT/IoHT communication environments including general security and privacy criteria as required from other networks (i.e., smart grid and wireless sensor networks) [25, 26].

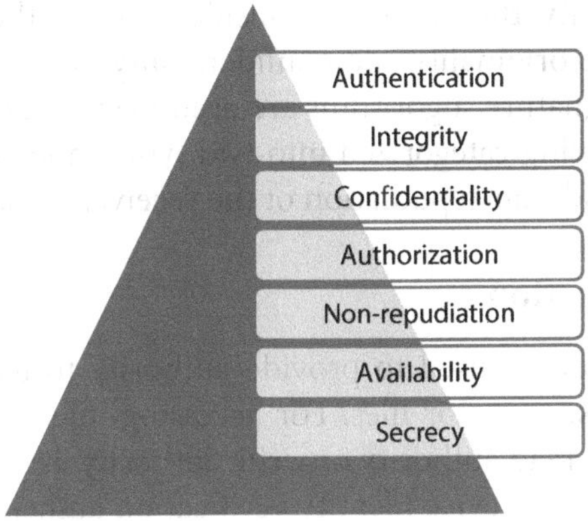

Figure 4.3 IoT issues and challenges.

4.3.1 Authentication

The authentication feature validates the identity (identification of communicators) or the communications exchange (message authentication). The identity will be verified simultaneously by both the sender and recipient before safe communication begins. It needs a variety of entities, such as IoT computers, diverse servers (e.g. cloud/fog servers), separate users (i.e. mobile/static), cloud service providers, and gateways, which require authentication of each other, depending on IoT deployment.

4.3.2 Integrity

Integrity means maintaining the fairness and authenticity of the proof. This guarantees that the message received during communication is not wrongly inserted, unnecessary deletion, and modification. We must defend the data from any form of malicious handling.

4.3.3 Confidentiality

Privacy guarantees privacy security from unauthorized individuals being accessed. Often known as anonymity, messages exchanged on the channel will thus be secured from any kind of attack on the sharing of information.

4.3.4 Non-Repudiation

Non-repudiation is a security feature in which the sender doesn't deny that he is sending data. No one can challenge the authenticity of the sender. In information security, this feature is considered to be the most important which ensures the originality and confidentiality of the original message. For this digital signature, algorithms are used to ensure the authenticity of the user. This is being categorized into two types: non-repudiation of the sender (origin) and non-repudiation of the receiver (destination).

4.3.5 Authorization

This security feature is used to provide authority to users or customers to access data, resources, or files. For accessing, file authorization levels are decided either read authority or write authority. It is normally accompanied by a device or device identity verification authentication protocol. Access policies are typically provided by a system and customer services authority (i.e., system administrator).

4.3.6 Availability

The availability of the content assures that it is only available to accepted parties.

4.3.7 Forward Secrecy

In forward secrecy, no IoT devices can access the data that are not available yet.

4.3.8 Backward Secrecy

In backward secrecy, no IoT devices can access the data that has been shared previously.

4.4 Internet of Healthcare Things (IoHT): Architecture and Challenges

From ancient times doctors have been symbolized as magicians. Their devotion and care bring healing, comfort, and hope to their patients. But as we know, before the enhancement of the technologies like the Internet of Things, audio, video and text communication and the latest gadgets, the visits and interactions with doctors are milestones for the patients to get the good treatment and suitable needy things on time. Furthermore, there were no available super-specialty hospitals nearby which patients can get themselves monitored and cured quickly and effectively [27].

In the healthcare sector, remote monitoring made the patients safe and healthy by unleashing the potential results in the empowering of physicians to deliver flawless care. Because of easy interaction with doctors; satisfaction in the minds of the patients has also been recorded as increasing simultaneously. As we know, IoT has put a drastic impact on the healthcare department by reducing healthcare costs and by improving treatment outcomes which results in the reduction of patients to stay in hospitals and of re-admissions.

4.4.1 IoHT Support

IoT is the modernization of healthcare by reshaping devices and increasing communication by delivering solutions to problems. IoT has numerous numbers of applications that will not only help patients and families but also provide significant support to hospital crews and insurance companies.

IoT support for Patients: The present world is full of automation. This automation brings certain changes in the lives of the people like one can measure one's own blood pressure, heart rate monitoring cuffs, glucometer, etc. just by wearing the bands that can detect all the above stuff. These devices can provide personalized attention that can be used to remind calorie count, daily exercise count, blood pressure variations, and much more. The evolution of IoT has changed people's lives, especially elderly patients, by continuous monitoring of health conditions. This makes the lives of individuals living alone very easy and relaxed. Moreover, if there is any change in a daily routine activity then these devices send messages to the doctor and patients' families to look forward to their health in a timely manner [28].

IoT for Physicians: With the help of wearable devices equipped with IoT, physicians got the trick to track the health of the patients more effectively without having a visit to them timely. Moreover, it makes the treatment plan and immediate medical attention up to the mark just by tracking. The best treatment given by the physician to their patient is by collecting the data from the IoT devices and with it, they can reach their desired outcomes.

IoT for Hospitals: The framework can be helpful in myriad regions while considering the healthcare platform. The possibility of keeping track of certain equipment and hardware such as nebulizers, defibrillators, or oxygen pumps that is even inclusive of the monitoring medical devices as well can be easily achieved by deploying the IoT based smart sensor technology. It also gives us the important feature of deployment of medical staff at a different location so that a better utility of management can be established. The hospitals and their environment are often subjected to certain infections which are the most important concerns for medical professionals. However, deep monitoring and IoT surveillance can avoid such instances of patients suffering to a great extent. Certain areas that are monitored regularly can be easily traced down with the help of this technology that included temperature monitoring, keeping a record of environmental changes as well as humidity percentage recording.

IoT for Health insurance companies: With the help of IoT intelligent devices plenty of opportunities are there for health insurers. To maintain their regular curricula, like claim and underwriting work, insurance companies also used a similar way by capturing the data that ensures them for getting being cheated by frauds and other outsiders' attacks. To bring transparency between the customers and the insurers IoT devices are also being used to provide better services like risk assessment processes, claim handling techniques, pricing, and much more. The data collected by IoT bring visibility to every decision which is laid by the insurance companies and their process outcomes. Customers are sharing and using the health data which is generated by IoT devices intelligently. The customers also received awards from the insurance companies for using IoT for their regular activities and treatment plans for precautionary health services. These ways help the insurers to keep the monitoring of customers and thoroughly validate their claims by the IoT devices using data capturing technique.

4.4.2 IoHT Architecture and Data Processing Stages

The various levels are being inclusive of the first sensor-based level, second co-ordinator-based level, and finally, the medical server level that has been illustrated in Figure 4.4. The illustrated architecture has been employed in the recently developed IoHT health monitoring systems [29, 30].

Medical equipment and controls are mounted on the sensor floor, which finally concludes the formation of a local network that can also be referred to as BSN an abbreviation for the Body Sensor Network. The sensors such as the RFID, Near-Field Communications (NFC) as well as BLE are being utilized in the sensor as well as server level for making the remote interchanges possible that are included in the Low-performance wireless infrastructure standards. The BLE secures the performance in a range of several network topologies that also include the mesh-based infrastructure or the star architecture. Also, the ultra-low power-based device to device communications that are required within implantable devices are being supported by the NFC and RFID [31]. The entire data as collected by the devices are being routed to the local devices such as tablets, notepads, or more

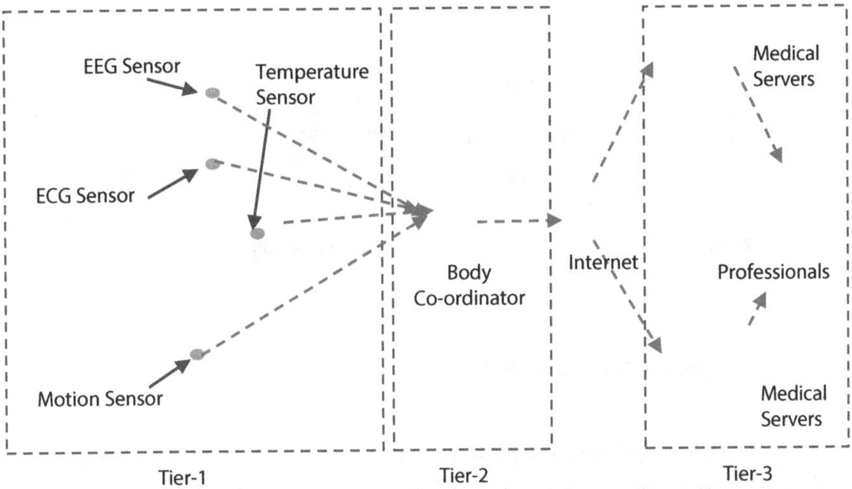

Figure 4.4 Architecture of IoHT.

complicated servers such as routers and gateways. Personal servers tend to process the patient data locally and store it to central medical servers before distribution. In case there have been certain issues associated with the network connection to the servers there should be provision to provide operation making use of the personal desktops. Healthcare practitioners, including doctors, will track patients' data online and provide patients with appropriate advice. In medical libraries with the consent of patients, early diagnostical algorithms and computer software and performance tests may be carried out as well. In the last decades, several IoHT-based healthcare systems for patient tracking have been proposed, but many of them do not provide protections for security and privacy, like MobiCare, or leaving them out as future work. There has been more focus on the problems and consequences forming associations with the use of power as well as usability when compared to the security and privacy-related problems or the protection of the data for the patients will [32].

IoT is a technology with four stages. All this information is so related that information is retrieved at one node and a value is generated at another point. Asset aggregation provides diverse sensations and possibilities for the everyday market (Figure 4.5).

Step 1: In this step several constituted devices like sensors, actuators, monitors, detectors, camera systems help in the collection of the data.

Step 2: The data which we get from step 1 is then undergone for further processing of data. In this step analog data is converted into digital data.

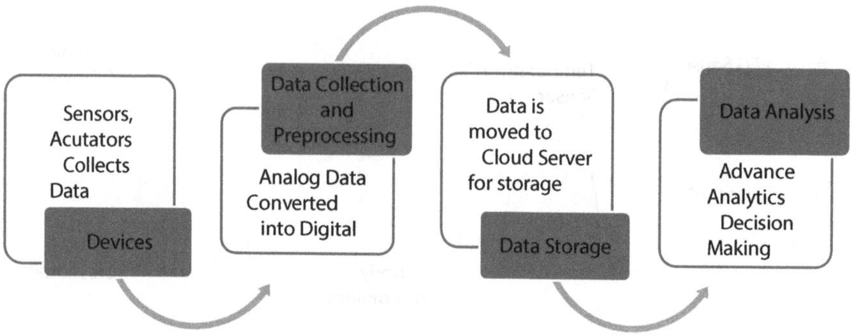

Figure 4.5 Stages of IoHT data analysis.

Step 3: This step involves pre-processing of the data that has been digitalized, then it undergoes standardization, and finally after the task has been accomplished the data is sent to the cloud or the data centers.

Step 4: The step performs the end task of analyzing and comprehending the data collected by making use of certain specialized Analytics programs that shall pave way for an efficient decision-making process.

The healthcare sector has marked an improvement in the medical services all in terms of the clinical visits reduction as well as enhancement in the work progress by bringing them on the platform IoTs. The remarkable improvement has been seen in the experience of the patient while accessing the medical care along with enhancement in the therapy outcomes.

4.4.3 Benefits Associated With Healthcare Based on the IoT

The most profound advantages associated with the IoT based Healthcare systems are being mentioned below:

- Cost Reduction: The decline in the frequent visits for the consultancy to the doctors along with admission in the hospital and bring a change in the cost invested.
- Improved Treatment: The physician got the full transparent information about the healthcare of the patient which helps to give the best treatment.
- Faster Disease Diagnosis: With help of continuous monitoring of the healthcare of a person can help to detect the disease at the early stages which helps to the diagnosis of that disease.

- Proactive Treatment: Proactive medical treatment is given to the patient with continuous health monitoring.
- Drugs and Equipment Management: In the healthcare industry the handling of drugs and other medical equipment is one of the major challenges through the interconnected devices these things are managed and utilized efficiently which reduces the cost.
- Error Reduction: Data generated by the IoTs intelligence devices are not only provided ineffective support in decision making but at the same time they make us certain in the smooth flow of healthcare operations along with fewer numbers of errors.

4.4.4 Challenges Faced by IoHT

These devices hold a tremendous amount of personal and professional data which is giving a rise to data security concerns.

Crucial steps are being implemented in the concern of security because IoTs are examining the new sides of patient healthcare with real-time monitoring of health all the time. For healthcare stakeholders, this data is a goldmine that is used for the improving health of the patient and gets new revenue opportunities from it. It also says that be wholly prepared for the ill-effect of digital power it may occur as a disaster in this interconnected world.

- As we know privacy and security of data is the biggest challenge because data generated daily is in a huge amount.
- Due to overload, the data transfer rate is lower and medical devices can create network congestion this is only due to a lack of protocols.
- As we know technology advancement results in an increment of prices which is a big challenge for common people to afford them.

4.4.5 Needs in IoHT

- To make fruitful decisions for the sake of patients the information collected from IoT is used. Fully interactive technologies strengthen and improve treatment effectiveness.

- IoT technology signifies emergency treatment and makes emergency assistance more efficient.
- IoTs are improving the patient's comfort by calling a patient in an emergency to use prescribed medicines.
- The urgent cases can be discussed with specialists across by world by a physician with help of IoT.
- Various generalized sensors and different types of equipment help doctors to examine their patient's health along with his/her internal feelings with ease which cannot be discussed properly.

4.5 Security Protocols in IoHT

The section ensures a detailed description of certain protocols for securing the communication environment while utilizing IoT. The security issues as described in the above section can be handled by different protocols which are categorized in the following categories (Figure 4.6). These security protocols can also be used in the security model of the Internet of Healthcare Things (IoHT) [33].

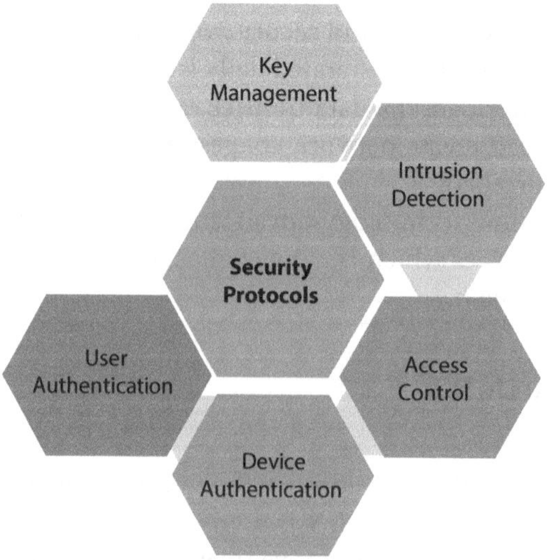

Figure 4.6 Security protocols in IoHT.

4.5.1 Key Management

A key management protocol addresses access, distribution, and implementation of encrypted keys between IoT/IoHT communication entities. The formulation of the entire process has been achieved by dividing the steps into certain portions for the performance. The steps are inclusive of the series of work that includes generating the key, key exchange, key usage as well as the key cancellation step. A key management mechanism, involving key server knowledge (i.e. trustworthy device entities), separate user sort (i.e. telephone, or static), and multiple machines, is using a cryptographic method (i.e., IoT devices). A strict key management mechanism should be developed to perform protected communications [34].

4.5.2 User/Device Authentication

It is a method for defining and checking the identity of contact groups. In shared authentication environments different users can share their identity (i.e. patients, an intelligent medical device). One can authenticate another user or device to access or participate in the communication. For future communication, they can create a session key that can be accessed by every member of the group.

4.5.3 Access Control/User Access Control

There has been a use of the Access Control whose main objective is to limit the resources on network/system for the user access controlling computer. Links to and access to several available resources have been provided to the user/device through this process. The studies brought about the conclusion of the association of smart devices with the system architecture of both IoT and IoHT that shall ensure the increased lifetime of the framework. There can be instances of machine breakdown arising due to the battery getting failed or some physical injuries. An opponent could also be able to deploy malicious devices on the network. The literature interpretation concludes the requirement of the differentiation of the systems that are original from those whose intentions are malicious. The objective can be achieved by designing of secure gateway for granting the controlling mechanisms that have the capability of segregating the systems from accessing the IoT/ToHT platform.

4.5.4 Intrusion Detection

Intrusion or anomaly detection is one of the biggest challenges over today's fast-growing communication system and internet services. As applications over the internet increases, it increases security threats. So, to detect such malicious activities over the internet, intrusion detection is considered to be an effective security tool. According to the study, it is seen that machine learning techniques are considered to be a more widely used application to design more efficient intrusion detection tool for achieving high accuracy rate as well as low false alarm rate. A more accurate predictive model can be built for such a large network dataset by applying a machine learning approach which is quite impossible with traditional security tools. The machine learning-based IDS learns the pattern of data packets captured in the network to determine its behavior to be an anomaly or normal. But traditional IDS system will only identify such attacks that are known to the machine. It cannot determine the unknown attack and its behaviour. There is the possibility of architecture interference by the botnets program that specifically targets the memory of the IoT architecture by bringing the malware into it via specific attacker coding. The result of such attacks is the system's complete breakdown or the malfunctioning of the program. Certain situations may lead to improper activity by the architecture. There is the possibility of certain more extreme and outrageous situations for instance the shock provided to the patient who has a pacemaker implanted inside him that can consequently lead to his/her demise. The conclusion being drawn, points towards ensuring proper protection in the IoT framework during communication from unnecessary intrusion [33–38].

4.6 Application of Machine Learning for Intrusion Detection in IoHT

Today Internet of Things (IoT) stands as a huge milestone in the era of technological advancement especially for the millennials given that this generation has only experienced the post-globalization period of the human timeline. When it comes to the healthcare monitoring department, manual inspection in the physical presence of a doctor or other qualified medical assistants was an inarguable necessity to monitor a patient's recovery. Medical test results were waited on for days which then had to be interpreted based on the diagnosis. The studies depict the more crucial and difficult security related issues along with the privacy factor in the system of healthcare being driven by the IoHT than in the systems which are dependent on the IoT platform for monitoring. The system prerequisites of the IoHT platform requires the additional device localization services in opposition to the IoT based working architecture for improvement in the performance in respect to privacy as well as security issues. The architectural design provides the level-wise function in the IoHT systems due to which the security and privacy guidelines have to be designed for each layer. The proper privacy designing has to be achieved for each layer and hence prevent it from the several network-based attacks that are intended to be precisely identified through these algorithms. Some of the major issues are being mentioned in Figure 4.7.

In this section, some major contributions of researchers are focused on the field of intrusion detection using machine learning or artificial intelligence. In [33] author designed a model for anomaly detection in a network for IoT applications. This model extracted a sample from data to analyze the behavior of the network traffic that can cause compromise to the network. In [39] the author provided a survey on different machine learning approaches for intrusion detection. The author also proposed a cloud-based IoT framework that applies machine learning tools to provide security and privacy to the data. In [40] the author used the features of biometrics along with machine learning to build a real-time healthcare monitoring system. The monitored data is remotely diagnosed and treatment decision is taken over a secure network. Man-in-the-middle attacks were considered in this work as an intrusion that was detected using machine learning and there was an improvement in detection rate from 7 to 25%. In [41] the author designed a model to secure the network in which medical devices are connected. An agent-based model was designed with the integration of regression machine learning to observe the anomaly sensor's data. In [42] the author developed a lightweight anomaly-based IDS to differentiate network traffic as an anomaly or normal. The author reduced the dimension of data by using principal component analysis. The cloud storage system was assumed for this work. In [43] the author proposed a deep learning intrusion detection system over an IoT fog network.

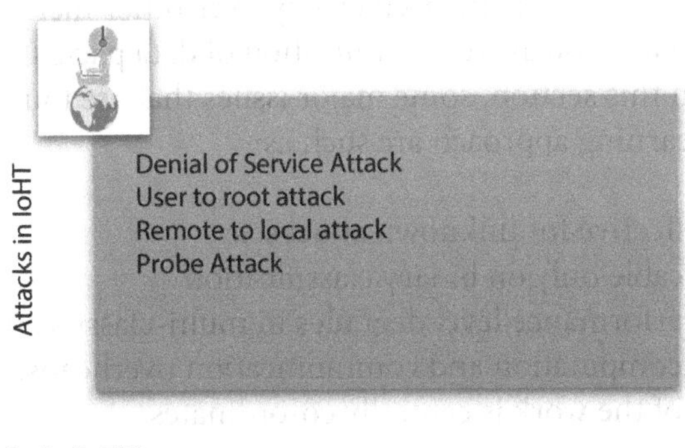

Figure 4.7 Attacks in IoHT.

The processing data is collected in such a distributed network with the application of fog nodes. The collected data is then processed and analyzed using deep learning. This work used the backpropagation learning that may cause complications while training the network. In [44], for faster training and accurate result, work was extended with OS-ELM machine learning algorithm in fog based IoT architecture. In [45], the author improved the work by applying an ensemble of online sequential extreme learning machine (EOS-ELM) for malicious activities detection for the fog–IoT network for healthcare applications. In [46] the author applied Deep Belief Network (DBN) for intrusion detection in IoT and achieved a precision rate of 99% for the known attack but for unknown attacks its precision rate was 86%. In [47] the author presented the PCA-GWO algorithm for intrusion detection in IoHT and achieved an accuracy of 99.9% and sensitivity of 93.2%. In [48] the author proposed a distributed intrusion detection for IoT applications in smart cities and achieved 99% of detection rate but for binary classification.

4.7 Proposed Framework

It is known that in the IoT environment, a large amount of data are produced and communicated over the network. To handle and organize such a huge amount of data over a network for analysis, accuracy is one of the important factors to be focused on. For a large dataset, manual feature selection takes a quite long time while communication among doctors and patients. In IoHT, several sensors are used to monitor the health of an individual. These sensors send sensitive data over an insecure network. So, it is required to monitor network traffic to prevent sensitive data to be compromised by intruders. This can be achieved by applying machine learning as well as deep learning. For network security, intrusion detection, deep learning proves their efficiency over other methods for feature selection and reduction as well classification of data packets as an anomaly and normal. In this section, some major issues that are found in the existing machine learning approach are such as:

- Not effective for unknown attacks.
- Applicable only on binary classification.
- The performance level degrades in multi-classification.
- High computation and communication overheads.
- Most of the work is centrally co-ordinates.

- If the central administrator is compromised then the decision may also alter.
- Irrelevant features identification increases computation time and reduces accuracy rate.

The proposed framework for intrusion detection in IoHT is illustrated in Figure 4.8. This framework gives a general overview of the framework that is composed of basically three layers: data layer, communication or network layer, storage, and analysis layer. The data layer is composed of smart sensor data or any information related to patients, hospitals, or insurance companies, etc. The collected data, either from sensors or any data user, is collectively communicated to the next layer i.e., communication or network layer. This layer is composed of gateway or switching devices that are responsible for analyzing the collected network data. The abnormal packets are analyzed using machine learning and reported to the administrator. The normal data packets are then transmitted to the next layer for their storage and analysis purposes. The collected data is analyzed

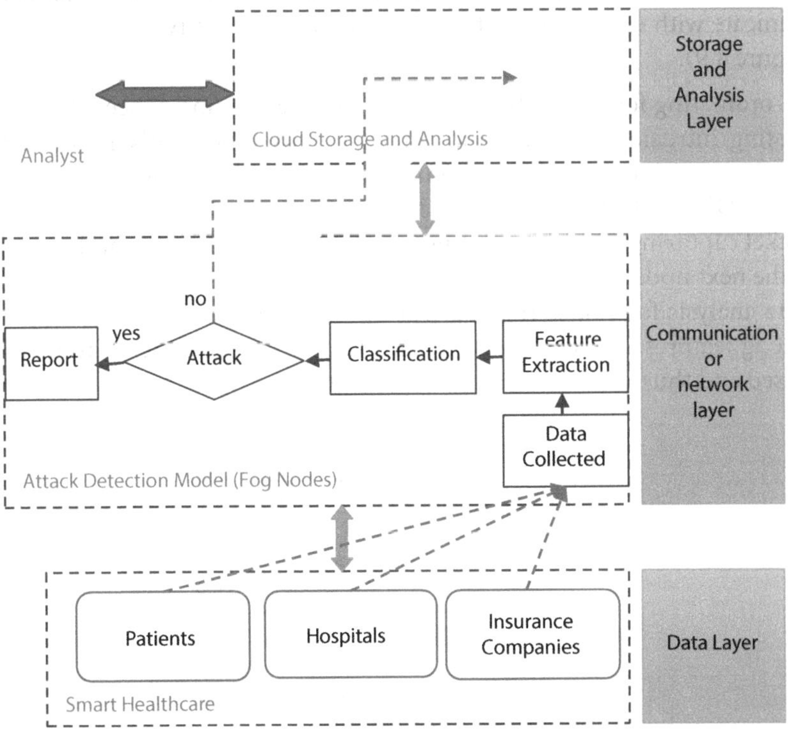

Figure 4.8 Security Protocols in IoHT.

by the doctors or specialists which are further stored at the cloud server from where patients can access the report by authenticating themselves. So, this framework is bidirectional which is secure, robust, and flexible.

The algorithm for the proposed model is illustrated below:

Algorithm:

1. Begin
2. Input: Intrusion Feature set={F_1, F_2,....F_n}
3. Output: Intrusion Report → Data(status)
4. Node initialization and deployment
5. Packet capturing fog←Sensor nodes
6. Feature Extraction
7. Data Analyst fog →Feature co-relation and reduction
8. Ensemble DNN training
9. Decision making fog →Data(status)
10. Generate alert
11. Update learning model
12. End

The proposed model consists of four distributed fog nodes that communicate with each other so that they can better identify network attacks (Figure 4.9).

Co-ordinating fog node: This node handles the training process from the existing intrusion database and trains the ensemble DNN network for creating decision rules. The training is performed based on protocol type TCP, UDP, and ICMP.

Packet capturing fog node: This node collects network traffic and passes it to the next node.

Data analysis fog node: This fog will extract features and find co-relation among features. The co-related features are further considered to be processed and thus results in feature reduction.

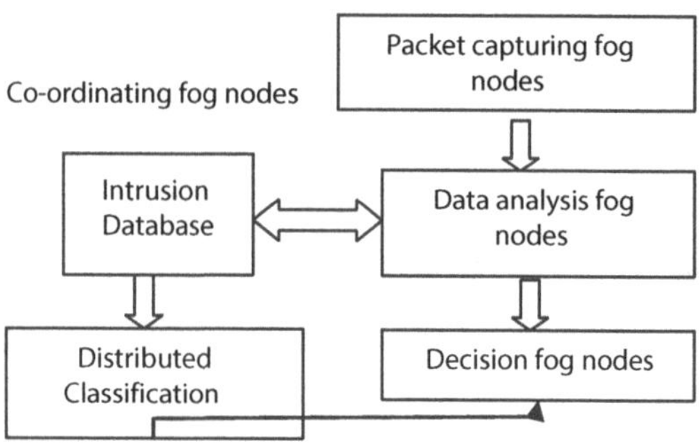

Figure 4.9 Proposed Multi-Fog architecture.

Decision fog node: This fog is responsible for deciding to analyze and process the uncategorized data packets. The data packets are classified according to protocols, i.e., TCP, UDP, and ICMP. It allows network and security administrators to perform reporting tasks on all detected intrusions by interacting with the intruder database and deep training algorithm. This fog node is distributed in nature which is illustrated in Figure 4.10.

The following key features of the proposed algorithm are discussed below that contributes in direction of future research:

- The proposed framework can enforce security and trustworthiness on the Internet of Healthcare Things (IoHT) by amalgamating privacy and a deep learning approach.
- The feature selection using bio-inspired optimization technique removes the drawback of a local minimum and they converge faster.
- By applying feature reduction along with a deep learning approach, training time is reduced.
- Highly related features are needed for improvement in performance level. Unnecessary features will cause calculation complexity.

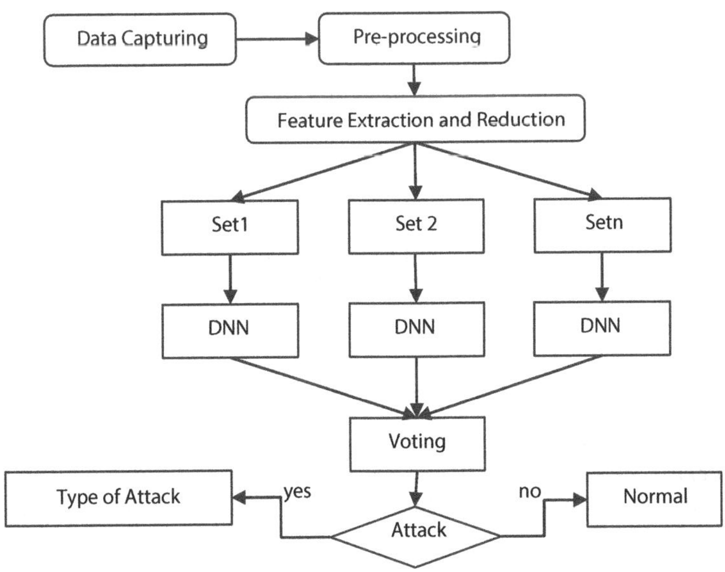

Figure 4.10 Proposed distributed classification.

- The proposed framework is effective in the IoHT scenario as the methodology can handle huge medical data along with the best features.
- Faster execution with reduced features results in a faster alert of intrusion and prevention measures can be applied accordingly.

4.8 Conclusion

Internet of Healthcare Things (IoHT) had emerged as an uprising technology for improvising the healthcare sector for fast and effective treatment of patients. But its deployment over the internet had raised security and privacy concerns. Several existing architectures, frameworks, algorithms, or protocols seem to be vulnerable to network attacks. So, the main scope of this chapter is to give an overview of the various intrusion detection system for IoT or IoHT architecture. But still, there are some limitations on existing techniques as every day new attacks are also introduced. So, the primary objective of this chapter is to propose a distributed framework to preserve the privacy of the data using deep learning such that the complexity of processing is reduced. The feasibility of the framework will incorporate privacy, security, accuracy, and reliability.

References

1. Gubbi, J., Buyya, R., Marusic, S., Palaniswami, M., Internet of Things (IoT): A vision, architectural elements, and future directions. *Futur. Gener. Comput. Syst.*, *29*, 7, 1645–1660, 2013.
2. Atzori, L., Iera, A., Morabito, G., The internet of things: A survey. *Comput. Netw.*, *54*, 15, 2787–2805, 2010.
3. Kumar, M., and Chand, S., A Secure and Efficient Cloud-Centric Internet-of-Medical-Things-Enabled Smart Healthcare System With Public Verifiability. *IEEE Internet Things J.*, *7*, 10, 10650–10659, 2020.
4. Song, T., Li, R., Mei, B., Yu, J., Xing, X., Cheng, X., A privacy preserving communication protocol for IoT applications in smart homes. *IEEE Internet Things J.*, *4*, 6, 1844–1852, 2017.
5. Tawalbeh, L., Muheidat, F., Tawalbeh, M., Quwaider, M., IoT privacy and security: Challenges and solutions. *Appl. Sci.*, 10, 4102, 2020, https://doi.org/10.3390/app10124102
6. Weber, R.H., Internet of Things—New security and privacy challenges. *Comput. Law Secur. Rev.*, *26*, 1, 23–30, 2010.

7. Ang, K.L. and Seng, J.K.P., Application Specific Internet of Things (ASIoTs): Taxonomy, applications, use case and future directions. *IEEE Access*, 7, 56577–56590, 2019.
8. Montori, F., Bedogni, L., Bononi, L., A collaborative Internet of Things architecture for smart cities and environmental monitoring. *IEEE Int. Things J.*, 5, 2, 592–605, 2018.
9. Khan, M.S., Islam, M.S., Deng, H., Design of a reconfigurable RFID sensing tag as a generic sensing platform toward the future Internet of Things. *IEEE Int. Things J.*, 1, 4, 300–310, 2014.
10. Tran-Dang, H., Krommenacker, N., Charpentier, P., Kim, D., Toward the Internet of Things for physical internet: Perspectives and challenges. *IEEE Int. Things J.*, 7, 6, 4711–4736, 2020.
11. Li, X., Wei, P., Wei, Z.J., Guosong, L., Ping, W., Research on security issues of military Internet of Things. *Int. Comput. Conf. Wavelet Active Media Tech. Inf. Proc. (ICCWAMTIP)*, pp. 399–403, 2020.
12. Guth, J., Breitenbücher, U., Falkenthal, M., Leymann, F., Reinfurt, L., Comparison of IoT platform architectures: A field study based on a reference architecture. *Cloudification of the Internet of Things (CIoT); Paris, France*, pp. 1–6, 23–25 Nov 2016.
13. Bansal, M. and Gandhi, B., IoT based development boards for smart healthcare applications, in: *International Conference on Computing Communication and Automation (ICCCA)*, Greater Noida, India, pp. 1–7, 2018.
14. Arias, O., Wurm, J., Hoang, K., Jin, Y., Privacy and security in internet of things and wearable devices. *IEEE Trans. Multi-Scale Comput. Syst.*, *1*, 2, 99–109, 2015, https://doi.org/10.1109/TMSCS.2015.2498605.
15. Yang, G., Jiang, M., Ouyang, W., Ji, G., Xie, H., Rahmani, A.M., Liljeberg, P., Tenhunen, H., IoT-based remote pain monitoring system: From device to cloud platform. *IEEE J. Biomed. Health Inf.*, *22*, 6, 1711–1719, 2017, https://doi.org/10.1109/JBHI.2017.2776351.
16. Das, A.K., Wazid, M., Kumar, N., Khan, M.K., Choo, K.K.R., Park, Y., Design of secure and lightweight authentication protocol for wearable devices environment. *IEEE J. Biomed. Health Inf.*, *22*, 4, 1310–1322, 2017, https://doi.org/10.1109/JBHI.2017.2753464.
17. Kumar, P., Braeken, A., Gurtov, A., Iinatti, J., Ha, P.H., Anonymous secure framework in connected smart home environments. *IEEE Trans. Inf. Forensics Secur.*, *12*, 4, 968–979, 2017, https://doi.org/10.1109/TIFS.2016.2647225.
18. Wazid, M., Das, A.K., Kumar, N., Conti, M., and Vasilakos, A.V., A novel authentication and key agreement scheme for implantable medical devices deployment. *IEEE J. Biomed. Health Inf.*, *22*, 4, 1299–1309, 2017.
19. Dua, A., Kumar, N., Das, A.K., Susilo, W., Secure message communication protocol among vehicles in smart city. *IEEE Trans. Veh. Technol.*, *67*, 5, 4359–4373, 2017, https://doi.org/10.1109/TVT.2017.2780183.
20. Tripathy, A.K., Tripathy, P.K., Ray, N.K., Mohanty, S.P., iTour: The future of smart tourism: An IoT framework for the independent mobility of tourists

in smart cities. *IEEE Consumer Electron. Mag.*, *7*, 3, 32–37, 2018, https://doi.org/10.1109/MCE.2018.2797758.
21. Alotaibi, S.S., Registration center based user authentication scheme for smart e-governance applications in smart cities. *IEEE Access*, *7*, 5819–5833, 2018.
22. Farooq, M.S., Riaz, S., Abid, A., Abid, K., Naeem, M.A., A survey on the role of IoT in agriculture for the implementation of smart farming. *IEEE Access*, 7, 156237–156271, 2019.
23. Elijah, O., Rahman, T.A., Orikumhi, I., Leow, C.Y., Hindia, M.N., An overview of Internet of Things (IoT) and data analytics in agriculture: Benefits and challenges. *IEEE Internet Things J.*, *5*, 5, 3758–3773, 2018.
24. Das, A.K., Wazid, M., Kumar, N., Vasilakos, A.V., Rodrigues, J.J., Biometrics-based privacy-preserving user authentication scheme for cloud-based industrial Internet of Things deployment. *IEEE Internet Things J.*, *5*, 6, 4900–4913, 2018.
25. Das, A.K., Zeadally, S., He, D., Taxonomy and analysis of security protocols for Internet of Things. *Futur. Gener. Comput. Syst.*, *89*, 110–125, 2018.
26. Yeh, K.H., A secure IoT-based healthcare system with body sensor networks. *IEEE Access*, *4*, 10288–10299, 2016.
27. Khowaja, S.A., Prabono, A.G., Setiawan, F., Yahya, B.N., Lee, S.L., Contextual activity based Healthcare Internet of Things, Services, and People (HIoTSP): An architectural framework for healthcare monitoring using wearable sensors. *Comput. Netw.*, *145*, 190–206, 2018, https://doi.org/10.1016/j.comnet.2018.09.003.
28. Firouzi, F., Rahmani, A., Mankodiya, K., Badaroglu, M., Merrett, G., Wong, P., Farahani, B., Internet-of-Things and big data for smarter healthcare: From device to architecture, applications and analytics. *Future Gener. Comput. Syst.*, 78, Part 2, 583–586, 2018.
29. Aghdam, Z.N., Rahmani, A.M., Hosseinzadeh, M., The Role of the Internet of Things in Healthcare: Future Trends and Challenges. *Comput. Methods Programs Biomed.*, *199*, 105903, 2020, https://doi.org/10.1016/j.cmpb.2020.105903.
30. Latif, S., Idrees, Z., Ahmad, J., Zheng, L., Zou, Z., A blockchain-based architecture for secure and trustworthy operations in the industrial Internet of Things. *J. Ind. Inf. Integr.*, *21*, 100190, 2021, https://doi.org/10.1016/j.jii.2020.100190.
31. Yang, G.Z. (Ed.), *Body Sensor Networks*, vol. 1, Springer, London, 2006.
32. Chakravorty, R., A programmable service architecture for mobile medical care, in: *Fourth Annual IEEE International Conference on Pervasive Computing and Communications Workshops (PERCOMW'06)*, IEEE, p. 5, 2006, March.
33. Meidan, Y., Bohadana, M., Mathov, Y., Mirsky, Y., Shabtai, A., Breitenbacher, D., Elovici, Y., N-BaIoT—Network-based detection of IoT botnet attacks using deep autoencoders. *IEEE Pervasive Comput.*, *17*, 3, 12–22, 2018.

34. Kao, Y.W., Huang, K.Y., Gu, H.Z., Yuan, S.M., uCloud: A user-centric key management scheme for cloud data protection. *IET Inf. Secur.*, *7*, 2, 144–154, 2013.
35. Wazid, M., Das, A.K., Kumari, S., Khan, M.K., Design of sinkhole node detection mechanism for hierarchical wireless sensor networks. *Secur. Commun. Netw.*, *9*, 17, 4596–4614, 2016.
36. Wazid, M. and Das, A.K., A secure group-based blackhole node detection scheme for hierarchical wireless sensor networks. *Wirel. Pers. Commun.*, *94*, 3, 1165–1191, 2017.
37. Wazid, M. and Das, A.K., An efficient hybrid anomaly detection scheme using K-means clustering for wireless sensor networks. *Wireless Pers. Commun.*, *90*, 4, 1971–2000, 2016.
38. Wazid, M., Reshma Dsouza, P., Das, A.K., Bhat, K.V., Kumar, N., Rodrigues, J.J., RAD-EI: A routing attack detection scheme for edge-based Internet of Things environment. *Int. J. Commun. Syst.*, *32*, 15, e4024, 2019.
39. Ghosal, P., Das, D., Das, I., Extensive survey on cloud-based IoT-healthcare and security using machine learning, in: *2018 Fourth International Conference on Research in Computational Intelligence and Communication Networks (ICRCICN)*, IEEE, pp. 1–5, 2018, November.
40. Hady, A.A., Ghubaish, A., Salman, T., Unal, D., Jain, R., Intrusion detection system for healthcare systems using medical and network data: A comparison study. *IEEE Access*, *8*, 106576–106584, 2020.
41. Thamilarasu, G., Odesile, A., Hoang, A., An Intrusion Detection System for Internet of Medical Things. *IEEE Access*, *8*, 181560–181576, 2020.
42. Zhao, S., Li, W., Zia, T., Zomaya, A.Y., A dimension reduction model and classifier for anomaly-based intrusion detection in internet of things, in: *2017 IEEE 15th Intl Conf on Dependable, Autonomic and Secure Computing, 15th Intl Conf on Pervasive Intelligence and Computing, 3rd Intl Conf on Big Data Intelligence and Computing and Cyber Science and Technology Congress (DASC/PiCom/DataCom/CyberSciTech)*, IEEE, pp. 836–843, 2017, November.
43. Diro, A.A. and Chilamkurti, N., Distributed attack detection scheme using deep learning approach for Internet of Things. *Future Gener. Comput. Syst.*, *82*, 761–768, 2018.
44. Prabavathy, S., Sundarakantham, K., Shalinie, S.M., Design of cognitive fog computing for intrusion detection in Internet of Things. *J. Commun. Netw.*, *20*, 3, 291–298, 2018.
45. Alrashdi, I., Alqazzaz, A., Alharthi, R., Aloufi, E., Zohdy, M.A., Ming, H., FBAD: Fog-based attack detection for IoT healthcare in smart cities, in: *2019 IEEE 10th Annual Ubiquitous Computing, Electronics & Mobile Communication Conference (UEMCON)*, IEEE, pp. 0515–0522, 2019, October.

46. Balakrishnan, N., Rajendran, A., Pelusi, D., Ponnusamy, V., Deep Belief Network enhanced intrusion detection system to prevent security breach in the Internet of Things. *Internet Things*, *14*, 100112, 2019.
47. RM, S.P., Maddikunta, P.K.R., Parimala, M., Koppu, S., Gadekallu, T.R., Chowdhary, C.L., Alazab, M., An effective feature engineering for DNN using hybrid PCA-GWO for intrusion detection in IoMT architecture. *Comput. Commun.*, *160*, 139–149, 2020.
48. Rahman, M.A., Asyhari, A.T., Leong, L.S., Satrya, G.B., Tao, M.H., Zolkipli, M.F., Scalable machine learning-based intrusion detection system for iot-enabled smart cities. *Sustain. Cities Soc.*, *61*, 102324, 2020.

5

Analyzing Recent Trends and Public Sentiment for Internet of Healthcare Things and Its Impact on Future Health Crisis

Upendra Dwivedi

Department of CSE, G.L. Bajaj Institute of Technology and Management, Greater Noida, India

Abstract

Internet of Things is becoming part of modern human life as smart devices are accepted and accessible to everyone. Smartphones are a common example that shows IoT devices' reachability. Digitization is the focus of businesses and governments to offer better services to individuals. The digital medium has been adopted as a platform for sharing information about topics ranging over various domains. Internet of Healthcare Things (IoHTs) plays significant importance in offering medical services to a broader population using digital infrastructures. The proposed work objective is to present a sentiment analysis and topic modeling based approach for early warning of a health crisis which can be integrated with the IoHTs framework and can be fruitful in assisting medical specialists. In this research work, information related to IoHTs is collected followed by dataset generation. Topic modeling is performed on the proposed IoHTs dataset for predicting trends in IoHTs domains. Sentiment analysis towards IoHTs applicability is evaluated to find the overall sentiment orientation of peoples. In the current pandemic situation, variation in sentiment orientation of users towards IoHTs system is evaluated and analysis is carried out to find the effectiveness of IoHTs systems.

***Keywords*:** Internet of Healthcare Things, topic modeling, sentiment analysis, VADER, latent dirichlet allocation

Email: ud1985@gmail.com

Kavita Sharma, Yogita Gigras, Vishnu Sharma, D. Jude Hemanth and Ramesh Chandra Poonia (eds.) Internet of Healthcare Things: Machine Learning for Security and Privacy, (95–112) © 2022 Scrivener Publishing LLC

5.1 Introduction

The digital platform is widely used to disseminate information among users from news publishers, government organization, non-government organization and individuals. Digital infrastructure is witnessing more interest and investment in this decade. Smart devices are associated with and offering services to everyone without their awareness. Most of the services which are offered online are using IoT based devices as backend [1]. IoHTs use properties of the Internet of Things (IoT) to provide healthcare services and touch human lives in regions that are unattended by basic healthcare services. Healthcare systems are playing an important role in making government policy, the financial health of an individual and the demands of a medical specialist. Individual and government are very concerned about existing health infrastructure. Every country is facing challenges in terms of unavailability of sufficient health infrastructure and the relative gap in comparison to demand and supply. Internet of Health Things can play an important role in providing support to existing health infrastructure by incorporating more technology. It requires capital investment and time to establish health infrastructure which is not feasible for many countries. A government and health specialist can utilize the IoHTs to support the demand of healthcare systems.

The recent pandemic situations have created so much chaos in individual lives and propose challenges of inefficiency and scarcity of health infrastructures to almost every government. The pandemic has caused tremendous impact on every individual life all over the world. Every country has been affected by the frequently occurring outbreak of the pandemic and infectious disease. It puts tremendous pressure on existing health infrastructures which are already reeling under patient load and scarcity of medical practitioners. It has been felt that governments are not able to provide sufficient healthcare to individuals and lagging in providing healthcare services to everyone. An individual, especially from the economically poor sector, has been affected very badly. They are not in the situation to adopt standard healthcare practices. The world has introduced various measures to tackle the situation but all these are not efficient. During this pandemic, measures that are being adopted by the world have a tremendous impact on the individual from physical health as well as mental health [2, 3].

The digital media platform is used to aware people of health information in a health crisis. Online news portal has become a popular platform for sharing and tracking health information during the current pandemic. Healthcare topics are one of the trending topics among netizen on the various social media platform. Researchers have used online available information to predict the outbreak of infectious disease. The sentiment of the public can be analysed by analysing topics in trends related to health. Social media is witnessing an exponential increase in information related to health issues, possible treatment and government policy. Each information contains user perspective and state of mind which is crucial for analysing upcoming demands for healthcare infrastructures [4].

This research work is organized as follows: Section 5.2 presents a literature review of the research work of IoHTs. Section 5.3 provides an overview of IoHTs and the collection of data into the proposed dataset. Section 5.4 discusses the significant of IoHTs in numerous aspects and topic modeling is performed on proposed IoHTs dataset. Section 5.5 presents sentiment analysis on the proposed dataset followed by discussion and conclusion.

5.2 Literature Review

IoHTs systems are designed to support existing healthcare infrastructure. User's opinion has significantly mattered for the growth of innovative technology and solutions and plays an important role in technology future. It should be measured how a user's opinion is going to affect the evolution of upcoming technological trends. Healthcare services have always required the incorporation of innovation and technology. Authorities can take the help of these sentiment dynamics for better decision making and prioritize resources.

Kumar *et al.* have presented an overview of public opinion of IoT on social media. A comparative study is carried out to find the relevance between research on IoT and social media. A social media analytical framework is proposed to analyze data from the microblogging portal Twitter. Social media and IoT data are analyzed using bibliometric analysis and a social media analytics framework. The research article finding suggests that the public's interest is about the application of IoT while researchers are interested in the IoT technical aspect [1]. Zhou *et al.* have performed sentiment analysis of text data at local government areas (LGAs) data fetched from Twitter, a social microblogging platform. The study has been carried out in a pandemic period to evaluate the effect of COVID-19 on LGAs. The analysis shows that there is general sentiment positive polarity shifting to negative polarity with increased cases of COVID-19, political decision and social events [2].

Verma and Sapra have presented sentiment analysis for cardiovascular disease from text data collected from Twitter [3]. Medford *et al.* have presented an analysis on Twitter contents for the COVID-19 pandemic. The early stage of pandemic analysis is carried out to understand public beliefs and sentiments [4]. Syarif *et al.* have collected information from Twitter for creating a mental illness corpus. A rule based system is presented to analyze the level of depression as high, moderate and low using information fetched from social media. Negative sentiment lexicon corpus from publically available data is created for analyzing depression statement and a decision tree-based system is developed with linguistic style and expression sentiment features [5].

Khamparia *et al.* have presented a study on skin cancer growth from the last decades. Skin cancer detection in an early stage benefits users in diagnosis and treatments. A deep learning framework is proposed for the classification of cancer detection and classification with an integrated capability to the Internet of Health and Things (IoHT) framework. The proposed work utilizes the concept of transfer learning and VGG19, Inception V3, ResNet50, and SqueezeNet are selected as pre-trained architectures for feature extraction from images. Transfer learning is applied to IoHT based framework and performance evaluation shows low computational complexity and higher accuracy [6]. Talpada *et al.* have presented the study to analyze the correlation between numerous sentiment analysis methods: Valence Aware Dictionary and Sentiment Reasoner (VADER), TextBlob, Long Short Term Memory (LSTM) and Stanford CoreNLP. Domain-specific keywords affect sentiment orientation of the targeted text. A Twitter-based sentiment model is proposed to evaluate the influence of telemedicine for heart attacks and epilepsy. A manually annotated dataset of health related tweets is built. LSTM deep learning model provides better accuracy for large and evenly distributed dataset than VADER and TextBlob [7].

Carleen Hawn has presented an overview of the relationship between the healthcare system and social media. Social media has impacted the healthcare system by providing a paperless communication medium for patients and medical practitioner. "Hello Health" is a popular web-based application for providing medicine assistance to patients to patient support groups as well as large hospitals [8]. Lee *et al.* have presented ontology for health technology categorized into health technology and health information. The presented ontology is further subdivided into wearable technology, biotechnology, telemedicine, medical technology and mobile health. Ontology keywords are extracted from Google Trends, Web of Science and Hashtagify [9]. Grover *et al.* have presented a study of health technology from an analysis of Twitter data concerning people, diseases and technologies. The Twitter data is collected from 15,587 users and evaluated using a social media analytics framework through word analysis and hashtag analysis. The study suggests that users are using the Twitter platform for getting numerous disease information and present opinion for example acute, chronic and communicable diseases [10].

Amy *et al.* have presented a scoping review for health research data using the Twitter platform as a source of information. The review articles include electronic databases OvidEpub, Ovid MEDLINE (R) Daily and Ovid MEDLINE(R), Ovid Embase, Ovid PsycINFO, Ovid EBM, EBSCO CINAHL, Elsevier Scopus, and Thomas Reuters Web of Science Core Collection for developing a search strategy [11]. Bian *et al.* have presented an overview and evolution of the Internet of Things (IoT) using Twitter data sources and Google Trends. Sentiment analysis is carried out on public opinion on IoT and the result shows that an overall perception is positive among Twitter users. Authors have performed topic modeling to find topics on Twitter data related to IoT search. It shows a predominant discussion on business, technologies, privacy and security issues in IoT [12]. Social media offer research opportunity to researchers to get a better understanding of numerous domains. Health related issues are among the important issues where social media enhance data availability for various health problems. Authors have carried out statistical, open vocabulary and Linguistic Inquiry Word Count (LIWC) analysis for evaluation of health disorders mental illness post-traumatic stress disorder (PTSD), depression, bipolar disorder, and seasonal affective disorder (SAD) from data collected from user's opinion on Twitter social media. The correlations metrics are applied to analytics used in the research work and the result shows quantifiable signals for numerous mental disorders in the Twitter data [13].

Lee, Agrawal and Choudhary have proposed a flu surveillance system and used Twitter health data to monitor flu related symptoms in real time in the U.S. Twitter APIs are used to collect data related to flu symptoms followed through data pre-processing to clean the tweet data by removing unnecessary texts. The tweet is processed to extract text, tweet time and user locations. Flu similar activity is detected across the U.S. with geographical modeling. Text modeling is performed for flu-related comparison and interactive data visualization is done in support of the proposed system [14]. Paul and Dredze have introduced the extended Ailment Topic Aspect Model (ATAM) for discovering a range of diseases with their symptoms. ATAM is used to analyze millions of health-related tweets to monitor syndromic surveillance over time for varied geographic regions. The ATAM model outcome shows correlations between model output and available public health data [15].

5.3 Overview of the Internet of Healthcare Things

Technology has a significant role in the evolution of human beings and every decade witnesses new trends of technology completely changing human behavior and perception. Technology is also playing an important part in developing new technology in numerous domains for example, health informatics, biotechnology, e-commerce and logistics. The Internet is witnessing the growth of connected devices using the Internet of Things as predominant technology. IoT has been seen as a business opportunity by offering smart services to users with affordability. IoT represents a network of computing objects embedded with microprocessors to perform numerous tasks in our lives. IoT is the interconnection of devices having computation and communication capabilities. These smart devices are designed to transfer data to the backend network for performing intelligent operations. IoT based system can be made from machine learning, embedded system, real-time analytics and control systems [1, 9, 12].

In this research work, IoHTs data is extracted from numerous internet resources. The information is collected including opinions, news articles and expert review on IoHTs related system and applications. This information is selected due to its impact on broader audiences that are directly or indirectly affected. Government and non-government organizations weigh individuals' opinion for decision making. Technology adaptation needs time and the healthcare sector is also adopting IoT based technical solutions. Technology evolutions follow the view of general users and their expectation. The information which is collected for this research work includes user's need and expectation towards IoHTs technology. It is optimistic to assume that information someway affects evolutions of IoHTs based system applications.

The IoHTs dataset collects news information extracted from publically available information related to the topic 'Internet of Healthcare Things'. The recent information has more effect in finding technological trends and

	datetime	Media	Title	Article	Summary
0	2020-08-14 00:00:00.000000	TechRepublic	2025 forecast: Global IoT healthcare market lo...	A new report says the promising global IoT hea...	A new report says the promising global IoT hea...
1	2020-12-08 19:02:28.623799	ETF Trends	Healthcare x Internet of Things = A Recipe for...	Chatter about the Internet of Things (IoT) is ...	Chatter about the Internet of Things (IoT) is ...
2	2020-03-17 00:00:00.000000	IoT For All	Advantages and Disadvantages of Implementing I...	The market for IoT healthcare technology is fo...	The market for IoT healthcare technology is fo...
3	2020-06-30 00:00:00.000000	Embedded Computing Design	How IoT is Transforming the Healthcare Industry	The adoption of the Internet of Things (IoT) h...	In healthcare, the IoT is used for interconnec...
4	2020-01-16 00:00:00.000000	HealthTech Magazine	How the Internet of Medical Things Is Impactin...	What Is IoMT?\n\nThe IoMT is a connected infra...	The IoMT is a connected infrastructure of medi...

Figure 5.1 Sample of IoHTs news dataset.

user's opinion. So, IoHTs dataset includes information from January, 2020 to November 2020. The news information includes the time of information uploaded on internet, source of information, the title of news articles, the content of articles and summary of articles as shown in Figure 5.1. IoHTs news dataset includes 450 news articles from reputed online resources. This resource list of news information is listed in Table 5.1.

Table 5.1 News resources list considered for IoHTs news dataset.

News resources list considered for IoHTs news dataset			
TechRepublic	ETF Trends	IoT For All	Financial Express
Embedded Computing Design	HealthTech Magazine	Forbes	Tech Observer
MedTech Intelligence	IoT Business News	Computerworld	FutureIoT
ITProPortal	Security Intelligence	World Economic Forum	ETCIO.com
ETTelecom.com	PRNewswire	Economic Times	WRAL Tech Wire
TechCrunch	Business Insider	TechTalks	The Punch
GlobeNewswire	Business Wire	Livemint	HealthITAnalytics.com
The San Diego Union-Tribune	OpIndia	SecurityInfoWatch	ETHealthworld.com
Verizon Communications	EnterpriseTalk	AZoSensors	Visual Capitalist
ZDNet	Technology Record	CXOToday.com	European Pharmaceutical Review
YourStory.com	builtin.com	China Briefing	Healthcare IT News
Hindustan Times	The Indian Express	Money Morning	HITInfrastructure.com

In Figure 5.2, a sample list of articles from IoHTs dataset is selected. This list includes the title name of articles as displayed by numerous online information resources and uses for extraction for the content of articles in IoHTs dataset.

In IoHTs dataset, articles are collected from numerous news resources. In Figure 5.3, a list of a few top resources is compared for their relative contributions. These resources are Financial Express, Healthcare IT News, HealthTech Magazine, Technology Record, The Courier, PRNewswire, CXOToday.com, TechRepublic, Motley Fool, World Economic Forum, Security Intelligence, ETCIO.com, Economic Times, Forbes and IoT Business News. These resources are listed concerning their articles contribution. IoT Business News is one of the online resources which can be considered for finding recent news about IoHTs.

```
                  Sample List of Articles included in IoHTs news dataset
Article 0 :- 2025 forecast: Global IoT healthcare market looks good—a $188.2 billion opportunity
Article 1 :- Healthcare x Internet of Things = A Recipe for Disruption
Article 2 :- Advantages and Disadvantages of Implementing IoT in Healthcare
Article 3 :- How IoT is Transforming the Healthcare Industry
Article 4 :- How the Internet of Medical Things Is Impacting Healthcare
Article 5 :- The 5 Biggest Internet Of Things (IoT) Trends In 2021 Everyone Must Get Ready For Now
Article 6 :- How The Internet Of Things Can Help Hospitals Cope With Coronavirus
Article 7 :- IoT: How Internet of Things is taking over human lives
Article 8 :- Artificial Intelligence to the Internet of Things: Technologies that defined the way we
lived in 2019
Article 9 :- How IoT is Reshaping Healthcare
Article 10 :- 5 Key Tips To Keep Your IoT (Internet of Things) Network At Home Private And Secure
```

Figure 5.2 Sample list of articles included in IoHTs news dataset.

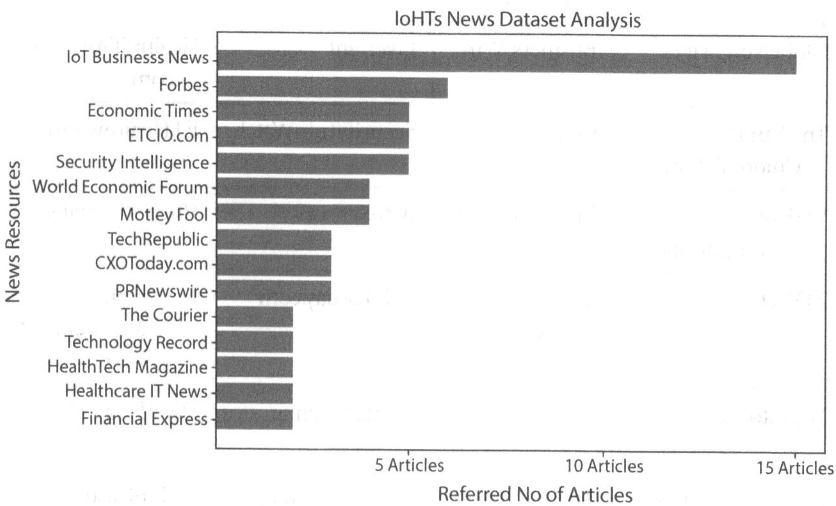

Figure 5.3 Referred no of articles from top news resources.

IoHTs dataset is analyzed for finding the most trending words which are commonly used online. In Figure 5.4, World Cloud is used to display word occurrence from a given dataset. These words show that the top words in the IoHTs dataset are Internet of Things, IoT, Healthcare, COVID 19, Market and Medical. IoHTs and COVID 19 word correlations show that IoHTs has a significant role in the recent COVID 19 pandemic.

News articles extracted from the internet include unwanted text, universal resource locator (URL), numbers, user handles, special characters and punctuation marks in the IoHTs dataset. In data pre-processing, the IoHTs database undergoes pre-processing removing unwanted characters and texts. The regular expression is applied to the original text from the dataset to remove unnecessary text '@' user handles, '#tags', numbers, universal resource locator (URL) and punctuation mark. In text information, the word is not in form of its root word. The context of words is important for performing topic modeling and sentiment analysis. The word context can be derived from using stemming and lemmatization in the field of Natural Language Processing (NLP). Stemming is used to reduce words into their root form but stemmed word may not be a valid word. Lemmatization transform words into their context and lemmatized word is the root form of the input word. In this work, lemmatization is performed on the IoHTs dataset to pre-process text into their context of words. The pre-processed dataset is passed to topic modeling and sentiment analysis for the Internet of Healthcare Things recent trends and its impact on the future health crisis.

Figure 5.4 IoHTs news dataset WorldCloud.

5.4 Performing Topic Modeling on IoHTs Dataset

Health monitoring is one of the challenging issues among government and non-government organizations. It requires human resources, materialistic resources and time at a large scale to implement public health monitoring at a large scale. Digital platforms and digital resources can provide solutions to develop a health monitoring system. A digital solution is cost-effective and resource-efficient. Data analysis is one of the possible solutions and can be used to develop a public health monitoring system. A digital health monitoring platform for all is one of the popular agenda among governments to provide healthcare solutions at low cost with access to everyone. Technology has facilitated digital solutions efficient and low-cost storage of health records of peoples. This information can be utilized for monitoring, education and decision making [2, 10].

Topic modeling algorithms are adopted to find thematic structure from given documents by word analysis. Annotation and labeling of texts are not required for performing topic modeling. In topic modeling, topics derived from given texts are the representation of words distribution. Latent Dirichlet Allocation (LDA), a probabilistic topic modeling algorithm is applied in this research work to find relevant topic towards the Internet of Healthcare Things. An LDA model is used to find words having similar observation and groups into topics from proposed IoHTs datasets [4]. News data from various sources is selected as the source of information to perform topic modeling to categorize given text data into latent topics. The dataset is pre-processed for removing unwanted information such as hyperlinks, Removing stop words and transform inflected words into their lemma using lemmatization. Pre-processed texts are passed to the LDA model applied a bag of words approaches on given texts to derive word vectors. The latent topic is probabilistic distribution representation over given texts [12, 16].

```
Performing Topic Modelling on IoHT Dataset
Topic #0: thing healthcare internet startup market nasscom innovation support healthineers siemens
Topic #1: internet thing iot healthcare way launch digital business risk care
Topic #2: thing internet medical iot healthcare home impact monitoring remote patient
Topic #3: healthcare iot billion market industry global artificial intelligence internet cio
Topic #4: internet iot thing hospital future help life coronavirus business make
Topic #5: iot covid device post health security play crucial society role
Topic #6: internet thing device healthcare smart case use iot new based
Topic #7: internet thing healthcare covid iot pandemic news coronavirus telecom future
Topic #8: healthcare smart security year cyber tech percent last organisation breach
Topic #9: thing internet healthcare iot technology global top player market news
```

Figure 5.5 Performing topic modeling on IoHT dataset.

Table 5.2 Possible abstract topic label derived from LDA topic model output.

Possible topic label	LDA topic model output
Internet of healthcare things start-up	Topic #0: thing healthcare internet startup market Nasscom innovation support healthineers siemens
Digital IoHTs business	Topic #1: internet thing IOT healthcare way launch digital business risk care
IoHTs impact on medical and home	Topic #2: thing internet medical IOT healthcare home impact monitoring remote patient
IoT healthcare and artificial intelligence	Topic #3: healthcare IOT billion market industry global artificial intelligence internet CIO
IoHTs future and Coronavirus	Topic #4: internet IOT thing hospital future help life coronavirus business make
IoT devices role in COVID 19	Topic #5: IOT covid device post health security play crucial society role
IoT devices and healthcare	Topic #6: internet thing device healthcare smart case use IOT new based
IoHTs and COVID 19 pandemic	Topic #7: internet thing healthcare covid IoT pandemic news coronavirus telecom future
Security in healthcare	Topic #8: healthcare smart security year cyber tech percent last organization breach
IoHTs technology in global market	Topic #9: thing internet healthcare IoT technology global top player market news

IoHTs dataset is fed to the LDA topic modeling and output is shown in Figure 5.5. These topics show what was trending in news related to IoHTs. The topmost trending topics are related to healthcare, start-up, innovation, hospital and COVID 19. The possible abstract topics are listed in Table 5.2 which are derived from the LDA model outcome. These abstract topics represent potential theme labels. Table 5.3 has listed LDA topic model output and corresponding possible topic labels. The top trending is Internet of healthcare things start-up, Digital IoHTs business, IoHTs impact on medical and home, IoT healthcare and artificial intelligence, IoHTs future and Coronavirus, IoT devices role in COVID 19, IoT devices and healthcare, IoHTs and COVID 19 pandemic, Security in healthcare

```
Performing Topic Modelling on IoHT Dataset during COIVD 19 period
Topic #0: healthcare thing internet iot industry intelligence artificial connected transforming lif
e
Topic #1: iot best approach case use apply way healthcare grow business
Topic #2: internet thing iot launch covid make home unveils market business
Topic #3: internet thing covid healthcare post telecom musk banking crucial society
Topic #4: iot covid healthcare pandemic news world lead telecom medical health
Topic #5: internet service thing risk cio news device growth help medical
Topic #6: iot future technology major prediction billion device adoptio medical speed
Topic #7: iot healthcare security market era covid thing internet coronavirus medical
Topic #8: iot thing internet healthcare help coronavirus life device smart billion
Topic #9: thing internet healthcare iot future startup innovation open support healthineers
```

Figure 5.6 Performing topic modeling on IoHT dataset during COIVD 19 period.

Table 5.3 Possible abstract topic label derived from LDA topic model output for the duration of COVID 19 pandemic.

Possible topic label	LDA topic model output for duration of COVID 19 pandemic
IoHTs and Artificial Intelligence	Topic #0: healthcare thing internet IOT industry intelligence artificial connected transforming life
IoHTs business	Topic #1: IOT best approach case use apply way healthcare grow business
IoT business in COVID 19	Topic #2: internet thing IOT launch covid make home unveils market business
IoT and COVID 19	Topic #3: internet thing covid healthcare post telecom musk banking crucial society
IoHTs and pandemic	Topic #4: IOT covid healthcare pandemic news world lead telecom medical health
IoT services and medical industries	Topic #5: internet service thing risk CIO news device growth help medical
IoT future technology	Topic #6: IOT future technology major prediction billion device adoptio medical speed
Secure IoT healthcare in COVID 19	Topic #7: IOT healthcare security market era covid thing internet coronavirus medical
IoHTs help life in coronavirus	Topic #8: IOT thing internet healthcare help coronavirus life device smart billion
Future start-up in IoHTs	Topic #9: thing internet healthcare IOT future startup innovation open support healthineers

and IoHTs technology in the global market. IoHTs and COVID 19 are highly discussed terms in news related to IoHTs. The topics 'IoHTs future and Coronavirus', 'IoT devices role in COVID 19' and 'IoHTs and COVID 19 pandemic' show that in a pandemic situation, IoHTs have played a significant role. The topic 'Security in healthcare' show that security in IoHTs is an important discussion and demands from all over the world.

LDA Topic modeling was applied on IoHTs news dataset for the duration of the COVID 19 pandemic. The news dataset was collected from March 2020 to November 2020 for analyzing correlation among IoHTs and COVID 19. Figure 5.6 shows the output of LDA topic modeling on IoHTs Dataset during the COVID 19 period. The topics are derived from the LDA topic modeling output. In Table 5.3, listed possible topic labels are derived from the LDA Topic Model Output for the duration of the COVID 19 pandemic. The topics 'IoT business in COVID 19', 'IoT and COVID 19, 'IoHTs and pandemic' 'IoT services and medical industries' and 'IoHTs help life in coronavirus' show that IoHTs are trending topics among the world and IoHTs have gained significant importance during the COVID 19 pandemic period.

5.5 Performing Sentiment Analysis on IoHTs Dataset

Sentiment analysis is well known as opinion mining is the field of Natural Language Processing (NLP). It is used to study subjective information from a given text. Lexical and sentiment-based VADER (Valence Aware Dictionary for sEntiment Reasoning), Text Blob and deep learning LSTM model are already adopted for analyzing sentiment orientation for numerous domains. VADER is a rule based sentiment analysis tool designed for sentiment analysis. Valence Aware Dictionary for sEntiment Reasoning (VADER) is a lexicon and rule based sentiment analysis tool for analyzing sentiments expressed in textual information [2, 7]. It emphasizes sentiment intensity using qualitative and quantitative methods and suitable for analyzing short text from the microblogging platform [5]. The objective of this proposal is to present a sentiment analysis based system for early warning of a health crisis that can be fully integrated with IoHT framework and can be fruitful in assisting medical specialists.

Sentiment analysis is performed on the sentiment orientation of IoHTs and their application to know how users are responding to recent developments. The recent pandemic has left a large impact on our existing health infrastructures which are not sufficient to handle the sudden massive spike in infectious or other diseases. Digital health infrastructure is the need of

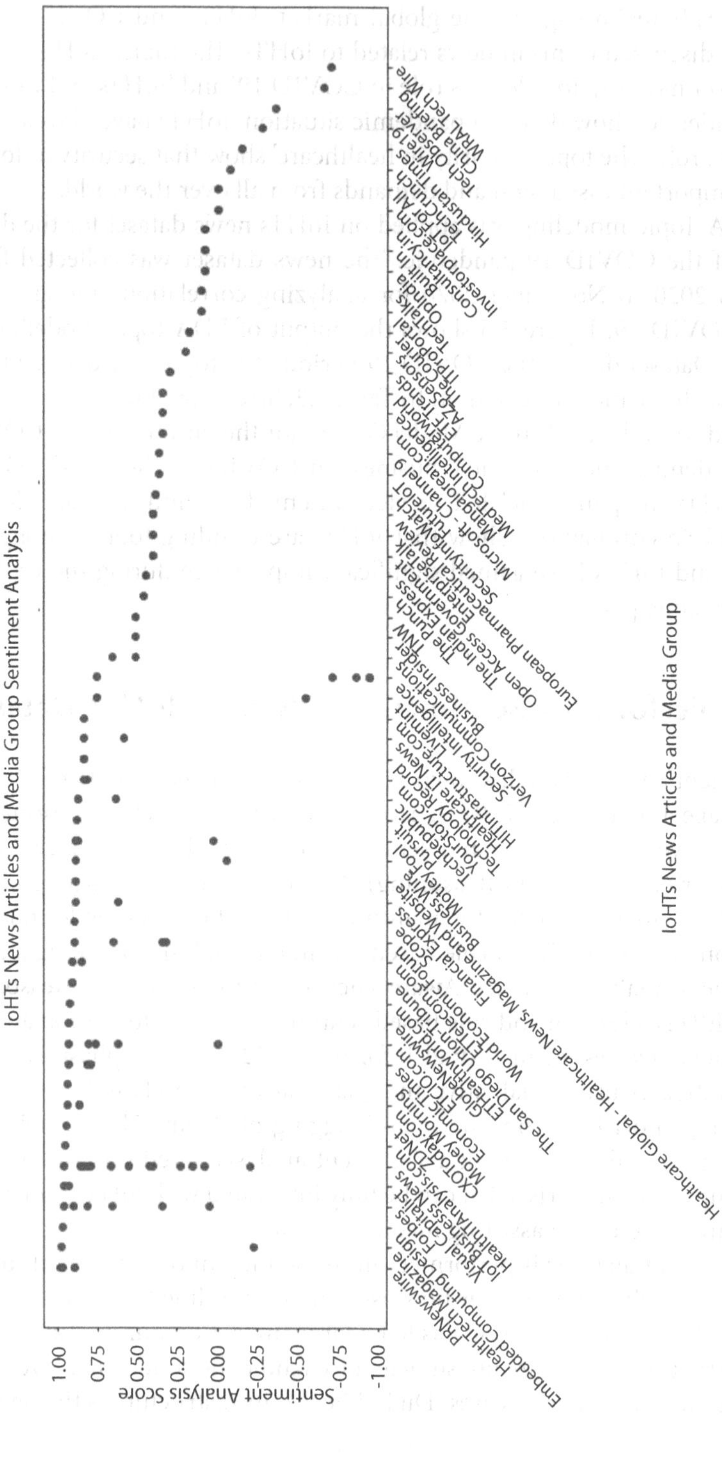

Figure 5.7 IoHTs News Articles and Media Group Sentiment Analysis.

the hour to support existing healthcare infrastructure. Technology adaptation proves to be fruitful for the existing system in numerous domains. IoT based solutions are already showing their impact on healthcare systems. These analyses help take effective measure in the prevention and cure of diseases outbreak. Some of the diseases that are research topics among researchers are influenza, Ebola Virus, Middle Eastern Respiratory Syndrome (MERS), Zika virus and many more [4].

Figure 5.7 represents a comparative sentiment analysis on IoHTs News Articles from numerous Media Groups. Most media groups have a positive sentiment orientation towards IoHTs and their applications. It shows a demand and a bright future for IoHTs and their applications. The top articles contributors are showing more positive sentiment on IoHTs discussion which shows that for the year 2020, IoHTs is a more promising technology.

IoHTs news dataset is analyzed for sentiment orientation for monthly news data extracted from numerous internet resources. This data is included from February 2020 to November 2020. Most of the news related to IoHTs news contains positivity as shown in Figure 5.8. Sentiment analysis score is a representation of sentiment orientation towards positivity or negativity of IoHTs news.

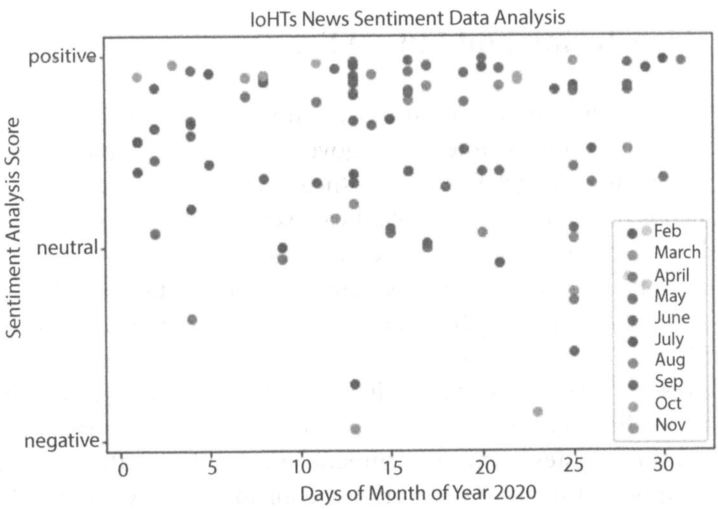

Figure 5.8 IoHTs News Sentiment Data Analysis for the year 2020.

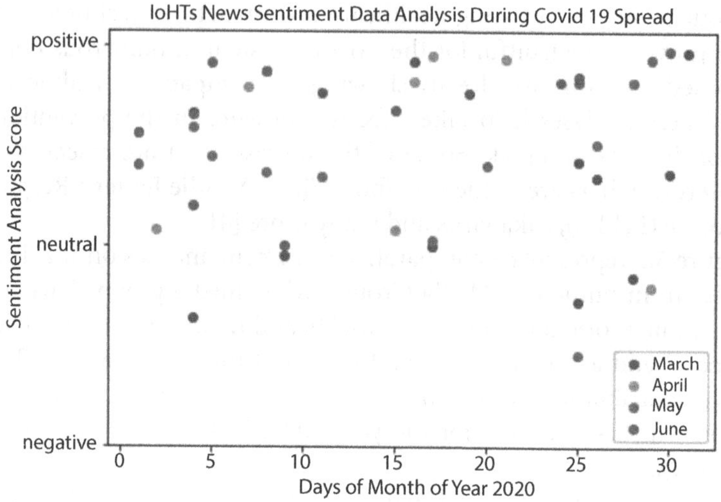

Figure 5.9 IoHTs News Sentiment Data Analysis during COVID 19 for the year 2020.

In Figure 5.9, IoHTs news data are analyzed from the period of COVID 19 spread duration from March 2020 to June 2020. News related to IoHTs mostly has positive sentiment in COVID 19. IoHTs' role in the pandemic duration is mostly optimistic and it plays an important part in supporting hospitality sectors in the pandemic duration as can be predicted from the topics modeling analysis.

5.6 Conclusion and Future Scope

The World Health Organization has discussed various health issues and disorders from time to time. Most governments are spending very low portions of their budget for the healthcare sector. It leads to the suffering of people from inaccessible healthcare services. High cost and scarcity of healthcare resources are some reasons for inefficient government programs. The pandemic has caused several economic issues as well as health issues. People are suffering from depression and anxiety similar to various mental health-related problems.

This research work has given a view on the dynamics of sentiment analysis and its effect on the Healthcare system, IoHT and its possible impact on the existing system with government intervention with different policy and support. The proposed work is a contribution towards providing a solution to offer effective utilization of IoHTs for numerous health issues.

An optimist approach can be helpful in a situation of an upcoming pandemic or disease outbreak. Governments can make better policies based on sentiment finding and can be able to provide healthcare resources to everyone who needs our country which is lacking so much infrastructure to support millions of people who cannot afford the healthcare system due to poor economic condition. Our approach can be beneficial to measure the effectiveness of the government policy with the IoHT effect. This proposal examines topic modeling and community sentiment in the situation of a recent health crisis.

In the pandemic era, resources require better utilization and management and this work can be useful while deciding IoHTs applications. In future work, the effect of health crisis and government policy on individuals and business will be analyzed. This research work shows the promising result and can be further extended by incorporating data collected from numerous survey methods with diversified data.

References

1. Kumar, R., Anand, A., Kumar, P., Kumar, R.K., Internet of Things and Social Media: A review of Literature and Validation from Twitter Analytics, in: *2020 International Conference on Emerging Smart Computing and Informatics (ESCI)*, IEEE, pp. 158–163, 2020, March.
2. Zhou, J., Yang, S., Xiao, C., Chen, F., Examination of community sentiment dynamics due to Covid-19 pandemic: A case study from Australia. 12185, SN Computer Science Journal, Springer, Australia, *arXiv preprint arXiv*:2006.12185, 2020, Available: http://arxiv.org/abs/2006.12185.
3. Verma, L. and Sapra, V., Semantic Analysis of Cardiovascular Disease Sentiment in Online Social Media, in: *Proceedings of International Conference on Advancements in Computing & Management (ICACM)*, 2019, October, Available at SSRN: https://ssrn.com/abstract=3462426or http://dx.doi.org/10.2139/ssrn.3462426.
4. Medford, R.J., Saleh, S.N., Sumarsono, A., Perl, T.M., Lehmann, C.U., An "infodemic": Leveraging high-volume Twitter data to understand early public sentiment for the coronavirus disease 2019 outbreak, in: *Open forum infectious diseases*, vol. 7, No. 7, p. ofaa258, Oxford University Press, US, 2020, July, https://doi.org/10.1093/ofid/ofaa258.
5. Syarif, I., Ningtias, N., Badriyah, T., Study on Mental Disorder Detection via Social Media Mining, in: *2019 4th International Conference on Computing, Communications and Security (ICCCS)*, pp. 1–6, IEEE, 2019, October.
6. Khamparia, A., Gupta, D., de Albuquerque, V.H.C., Sangaiah, A.K., Jhaveri, R.H., Supercomputing, Internet of health things-driven deep learning system

for detection and classification of cervical cells using transfer learning. *J. Supercomput.*, *76*, 1, 1–19, 2020.
7. Talpada, H., Halgamuge, M.N., Vinh, N.T.Q., An analysis on use of deep learning and lexical-semantic based sentiment analysis method on Twitter data to understand the demographic trend of telemedicine, in: *2019 11th International Conference on Knowledge and Systems Engineering (KSE)*, pp. 1–9, IEEE, 2019, October.
8. Hawn, C., Take two aspirin and tweet me in the morning: how Twitter, Facebook, and other social media are reshaping healthcare. *Health Affairs*, *28*, 2, 361–368, 2009.
9. Lee, J., Kim, J., Hong, Y.J., Piao, M., Byun, A., Song, H., Lee, H.S., Health information technology trends in social media: Using Twitter data. *Healthcare Inf. Res.*, *25*, 2, 99, 2019.
10. Grover, P., Kar, A.K., Davies, G., "Technology enabled Health"—Insights from Twitter analytics with a socio-technical perspective. *Int. J. Inf. Manag.*, *43*, 85–97, 2018, https://doi.org/10.1016/j.ijinfomgt.2018.07.003.
11. Zhang, A.J., Albrecht, L., Scott, S.D., Using Twitter for data collection with healthcare consumers: A scoping review. *Int. J. Qual. Methods*, *17*, 1, 1609406917750782, 2018.
12. Bian, J., Yoshigoe, K., Hicks, A., Yuan, J., He, Z., Xie, M., Guo, Y., Prosperi, M., Salloum, R., Modave, F., Mining Twitter to assess the public perception of the "Internet of Things". *PloS One*, *11*, 7, e0158450, 2016, https://doi. org/10.1371/journal.pone.0158450.
13. Homan, C., Johar, R., Liu, T., Lytle, M., Silenzio, V., Alm, C.O., Toward macro-insights for suicide prevention: Analyzing fine-grained distress at scale, in: *Proceedings of the Workshop on Computational Linguistics and Clinical Psychology: From Linguistic Signal to Clinical Reality*, pp. 107–117, 2014, June.
14. Lee, K., Agrawal, A., Choudhary, A., Real-time digital flu surveillance using Twitter data, in: *The 2nd Workshop on Data Mining for Medicine and Healthcare*, 2013, May.
15. Paul, M. and Dredze, M., You are what you tweet: Analyzing Twitter for public health, in: *Proceedings of the International AAAI Conference on Web and Social Media*, vol. 5, No. 1, 2011, July.
16. Campbell, J.C., Hindle, A., Stroulia, E., Latent Dirichlet allocation: extracting topics from software engineering data, in: *The art and science of analyzing software data*, pp. 139–159, Morgan Kaufmann, USA, 2015, https://www.sciencedirect.com/book/9780124115194/the-art-and-science-of-analyzing-software-data#book-info.

6

Rise of Telemedicine in Healthcare Systems Using Machine Learning: A Key Discussion

Shaweta Sachdeva* **and Aleem Ali**†

Department of CSE, Glocal University Saharanpur, U.P., India

Abstract

Telemedicine is the use of technology to clarify various elements of medical care from a distance. Telemedicine is defined as the use of telecommunications and information technology to deliver healthcare data and administrations to patients who are located far away. The data is moved from one segment to the next via electronic signs. With the utilization of electronic imaging gadgets, it is getting perceived as implied too far off clinical consideration and conferences with various patients. Telemedicine incorporates EMR (Electronic Medical Records) of patients at one spot which can diminish the number of cases, from where specialists cannot get to patient's records operational. AI is generally utilized in different uses of Telemedicine. It is assuming an indispensable part in numerous applications like medical science, accounting, and insecurity. AI plays an important role to find designs from clinical information sources and gives incredible capacities to anticipate infections. Organizing medical services is a smart method to use media transmission technology, which has been used to address a wide range of health concerns in a variety of ways. Telemedicine administrations can provide remedial administrations, diagnostic tests, and preventive arrangements, as well as patient guidance. It is useful for correspondence between specialists, drug specialists, patients, and those who live in different parts of the world. Telemedicine administrations are progressively used by patients, clinicians, and organizations.

***Keywords*:** Telemedicine, telecardiology, teleradiology, telepharmacy, ICT, artificial intelligence, machine learning, healthcare

**Corresponding author*: sachshaweta16@gmail.com

†*Corresponding author*: aleem@theglocaluniversity.in

Kavita Sharma, Yogita Gigras, Vishnu Sharma, D. Jude Hemanth and Ramesh Chandra Poonia (eds.) Internet of Healthcare Things: Machine Learning for Security and Privacy, (113–130) © 2022 Scrivener Publishing LLC

6.1 Introduction

AI is a wide multi-disciplinary field that has its branches in various areas, variable-based math, information preparing, and utilized in information examination and so on, that makes it extreme to surrender with an exact definition. Machine Learning is probably the best technique to gather data in for computerized manner for preparing information. In Machine Learning, we are not guiding the machines where to look. It fills in as a tree and has various branches and sub-branches. AI has the following classifications like Supervised Learning, Semi-directed Learning, Unsupervised Learning, and Reinforcement Learning.

Telemedicine is the utilization of data and correspondence transformation to give medical care directing to people who are living remotely from the medical supplier. As opposed to being a solitary innovation, telemedicine is important for a more extensive interaction or chain of care [1]. Telemedicine can improve this chain of care and along these lines; it upgrades the quality and proficiency of medical services which are given to patients. Telemedicine is likewise expected to expand the reasonableness and equity of the dissemination of administrations, because the availability of wellbeing administrations, particularly in far-off zones, can be improved. The utilization of more important methodologies like telephone and fax is common; telemedicine applications progressively utilize the most recent advancements in Personal Computer and organization in PC and also used other best techniques i.e. ICT.

Before appropriation into routine use, any innovation must end up being better than the methodology that it is proposed to supplant, and it must be more practical than the alternative. Telemedicine is no special case for this standard. There are developing telemedicine frameworks that require an evaluation and the progressing assortment of applicable information for regulatory purposes before they can be considered for routine use for a huge scope. Moreover, appraisal of telemedicine applications is expected to help to buy and arranging choices to screen and change the utilization of the innovation when it is set up. Telemedicine is the utilization of electronic correspondence and data advancements to give or support clinical consideration a way off [2]. Tele-health is characterized as the utilization of electronic data and broadcast communications advancements to help significant distance clinical medical services, patient and expert well-being related schooling, general well-being, and well-being organization.

6.2 Types of Machine Learning

A) Supervised Machine Learning
This learning is used to prepare and calculate an activity that matches with the contributions to related yields (Figure 6.1). One foundation of the managed learning task is the arrangement issue.

B) Semi-Supervised Learning
It is used for recognizing the best classifier from the unlabeled and named data. By utilizing unlabeled data, it moves the elite of arrangement. The achievement of this learning thoroughly relies upon not many basic suppositions.

C) Unsupervised Learning
It is the technique used for recognizing the best classifier from each unlabeled and named data. By utilizing unlabeled data, it moves the elite of grouping. The achievement of this technique thoroughly relies upon not many basic presumptions.

D) Reinforcement Learning
It is the technique of Machine Learning which is used to taking suitable action to get a huge reward in a particular situation.

6.3 Telemedicine Advantages

One of the best advantages of telemedicine is that it is helpful for military staffs who arc rcsiding far away. They have irregularly arranged in regions of the world that telemedicine is the most ideal approach to investigate and treat them effectively. The reality of the situation is that telemedicine can go wherever as it is both adaptable and amazing. Unmistakably this comprehensively used practice can be incredibly compensating especially when you consider and offers huge clinical information to experts in various countries around the globe. We don't have any clinical issues out there today.

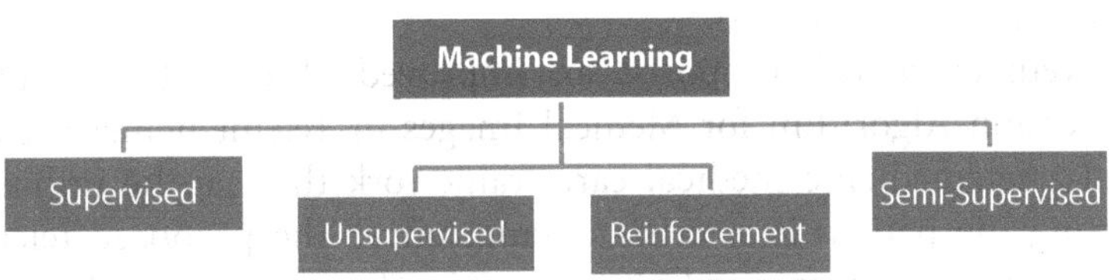

Figure 6.1 Types of machine learning.

The use of the new contaminations and clinical issues emerging is the best way to deal with learning faster with an ultimate objective to save lives or avoiding trouble for the patient to be treated.

6.4 Telemedicine Disadvantages

Telemedicine can be used regularly in various ways with laws and a set of rules which are applied before. It takes a lot of time to provide help anywhere to anyone. It becomes paid for them by Organizations and specialists ought to be settled as telemedicine transforms into a more natural practice. This can bring a problematic affirmation as patients have more clinical risk and are over-dependent on the telemedicine system. Due to these risks, patients who were depending on the usage of telemedicine concept and wrong information were recorded.

As we know telemedicine is not yet completely worked out, where it can be utilized carelessly or deftly. In any case, it gives a big benefit for the world and all its turns of events. There is no battling that the responsibilities it can make have limitless possibilities in any case. Extra time and efforts will be relied upon to arrange telemedicine for it to be a positively recognized treatment [3].

6.5 Review of Literature

Afza and Manjula [4], proposed the Eye-Opening Aspect of Telemedicine in Punjab have talked about the different specialized issues and their downsides in India. A bunch of norms should be executed. They have examined telemedicine regarding neuron careful crises and prescribed the norms to be continually refined for effective execution of this developing innovation.

Mohandass and Janet [5], proposed the Design and Implementation of a Telemedicine Network in the Sub Himalayan State of India that there are bunches of endeavors started by the private and the public area to improve the nature of clinical administrations. Advanced clinical libraries have additionally been presented.

Dwivedi *et al.* [6], proposed an Improved Three Pattern Huffman Compression Algorithm for Medical Images in Telemedicine; it characterized a telemedicine medical care framework that can be utilized for observing the patient in a crisis circumstance. The proposed telemedicine framework is comprised of a convenient/non-compact telemedicine area and a versatile/non-convenient base segment. The utilization of TCP/

IP makes the operability of frameworks in a telemedicine climate more proficient.

Kontaxakis *et al.* [7], proposed the Wireless telemedicine and m-well-being technologies, applications, and examination issues to clarify the utilization and future possibilities of telemedicine innovation with models. A concise conversation about the LINCOS project was additionally introduced. An itemized conversation on different specialized issues like pressure, man-made reasoning was introduced. Clinical sensors, home checking frameworks, and electrical medical records (EMRs) were used for the telemedicine framework.

Angaran [8] proposed the meaning of telemedicine, examining the telemedicine applications. Telemedicine application is utilized and recognized as the use of cutting edge correspondence advancements, the inside setting of clinical thriving, which offers care to patients. It connected with the two specialists and patients to get an ever-increasing number of favorable circumstances of this application.

Clifton *et al.* [9], proposed a two-way video-conferencing concept for the improvement of telemedicine application regarding remote dispensary, and joined with electronic gear, permitting trained professionals, and other similar thriving experts to supply complex medical services and exploit many patients who live miles missing from the area.

Traynor [10], proposed the perpetual telemedicine applications which contain constant consideration, getting ready, asking about the therapy, and opening prosperity to investigate, directly consider, send and get prosperity information, break down X-beams, and show prosperity specialists. Telemedicine applications were effectively taken care of with creative contraptions for medical services conveyance, medicines remedies, meds, and pictures of clinical sciences.

Conrath *et al.* [11], proposed the best way known as an edge of telemedicine called E-wellbeing, a later forward leap in telemedicine development analyzed in detail thereafter. Besides, a few fundamental E-wellbeing applications favor how telemedicine is significant for away from patients and accessible through any PC (for example PCs and compact workstation of PCs, PCs, and so on) associated with the Web.

Muller *et al.* [12], proposed the cost factor inside the local area, which demonstrates how telemedicine was introduced to the local area and for the implementation of this concept essential methods were used for medical care purposes. It shows the reality to focus on early things of telemedicine applications and it underlines different starting associations [13]. It stays vigorous and is usually used inside medical care applications and other near conditions where profits by these gadgets can, in any case, be seen.

McCue *et al.* [14], proposed the framework for far off patient checking utilizing devices like cell phones. It is produced for the provincial Indian populace who go through hypertension and hypotension. This framework is intended to catch pre-cardiovascular confined circumstances in individuals and is safely recommended for clinical/medical care. These frameworks utilized biomedical sensors and microprocessors for performing investigations and get treated.

Hayes *et al.* [15], proposed the framework for Telemedicine for Diabetics patients in rustic territories. As indicated by the aftereffects of the ICMR (Indian Council of Medical Research) scientists who assessed those 62.4 million individuals experiencing diabetes. Numerous endeavors have been accomplished for diabetic patients in metropolitan zones, while 70% of the populace lived in provincial zones. This percentage of assessment is expanded step by step because of the absence of mindfulness in individuals living in provincial zones, restricted admittance to medical care therapy because of transportation issues, and absence of prepared specialists in diabetes. The screening for diabetes is done in rustic territories which bring about undiscovered and inappropriate treatment to the patients.

The examination thought of a diabetes medical care model to focus on patients in a more moderate way utilizing a versatile van office having sorts of gears to analyze and in the event of a crisis they can impart to specialists through prepared professionals utilizing satellite correspondence utilizing VSAT. The examination presumed that the expense adequacy of medical care can be accomplished through telemedicine projects.

Harno [16] proposed an examination model, named thumping Telemedicine Technology for medical services experts utilizing diverse overview techniques to assess the utilization effect of different discrete organizations on telemedicine abilities in India. The ICT foundation in India somewhat moved from regular specialized segments to rapid fiber-optic broadband availability organizations. The examination additionally states it checked improvement, however, prompts progression and manageability in existing telemedicine abilities also.

6.6 Fundamental Key Components Needed to Begin Telemedicine

6.6.1 Collaboration Instruments

These instruments are the gadgets that allow the clinic to associate more patients with suppliers, for example, cell phones, tablets, and PCs.

6.6.2 Clinical Peripherals

Peripherals are used in telemedicine, like stethoscopes, ultrasound machines, or automated stethoscopes. Coupling these peripheral devices with different instruments and work measure programming enables the transport of predominant sound, video, pictures, and data in high quality for parental figures to give treatment and finding from various territories. These versatile peripherals are important to help with compatible devices to give comfort and ease.

6.6.3 Work Process

Work measure is the way to have full documentation for various subspecialties and modalities in different treatments. These provide complete information about patients at centers for treatment [17]. Similarly, it requires adequate programming to manage the complete pattern of partner patients to specialists and to involve telemedicine with their present IT resources, as Electronic Health Record structures.

6.6.4 Cloud-Based Administrations

Generally, telemedicine frameworks have run on big business frameworks. It is costlier to provide these medical services through one telemedicine system framework [18].

6.7 Types of Telemedicine

Telemedicine can be broken into three main categories.

6.7.1 Store-and-Forward Method

Telemedicine includes securing health information (like X-rays, CT scans, biosignals, and so on) and used to transmit health information to a specialist or clinical expert/doctors at a time which gives help to experts for evaluation when the patient was being disconnected (Figure 6.2). The presence of the two players is not required simultaneously. Dermatology, Radiology, and Pathology are regular treatments that are helpful for off-center telemedicine. The store-and-forward method is an appropriate method used to organize Electronic Medical Record in telemedicine structure ought to be a part of this exchange. It plays a vital role to store in-person information for actual assessment and store the history of the patient for giving

Figure 6.2 Store and forward method in telemedicine.

accurate treatment [19]. The store-and-forward interaction requires the experts/doctors to depend on a set of experiences reports and sound/video data in place of an actual assessment.

6.7.1.1 *Telecardiology*

ECG can be sent by utilizing phone and remote devices. Einthoven, the innovator of the ECG, really did tests with transmissions of ECG through phone lines first. This was because the emergency clinic didn't permit him to move patients outside the clinic to his research facility for testing of his new gadget. Teletransmission of ECG utilizes native strategies (Figure 6.3). One of the most seasoned known telecardiology frameworks (Teletransmission of ECG) was set up in Gwalior, India in 1975 at the GR Medical School utilizing various strategies without precedent for India. This framework was additionally used to screen the patients with pacemakers in remote areas. The control unit is done at the ICU in the hospital. This strategy assisted clinical with supporting reach in far-off territories [20].

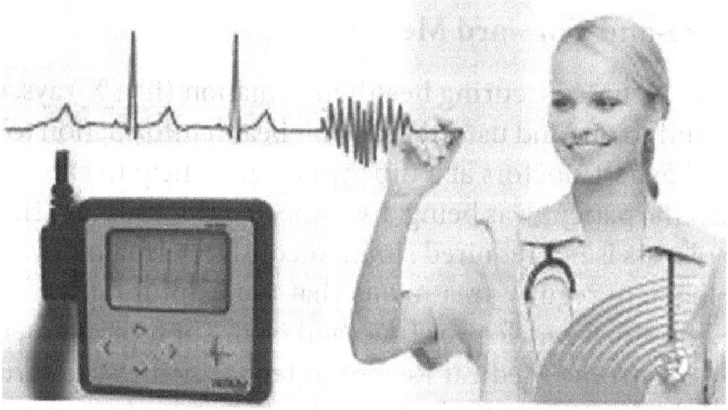

Figure 6.3 Telecardiology.

6.7.1.2 *Teleradiology*

Teleradiology is one of the techniques used to transmit radiographic pictures like X-ray beams, CT scans, and MRIs starting with one part then onto the next. For this interaction, three fundamental parts are required, a picture sending center, a transmission organization, and an accepting/picture audit center. The most average usage is two Personal Computers are connected through the Internet (Figure 6.4). The Personal Computer is a desirable device that should have a great presentation screen to visualize and to be cleared for health purposes. All the information to be accepted by Personal Computer will have a printer so that pictures can be printed for the next process. The teleradiology technique starts the process from the picture sending center. The radiographic picture, a modem or different devices are needed for the initial and last step of teleradiology. Then the picture is filtered and then sent through the organization associated with the getting Personal Computer [21].

6.7.1.3 *Telepsychiatry*

Teleps 1ychiatry, another part of telemedicine, likewise uses video conferencing for patients living in reserved regions to get to mental and physical treatment. It offers a wide scope of services to the patients, suppliers and experts, for example, counsel between the specialists, instructive clinical program, analysis and evaluation, drug treatment, and so on.

6.7.1.4 *Telepharmacy*

Telepharmacy is another developing pattern for giving medical and drug care to patients in faraway areas where they might not have actual contact with drug specialists/doctors. It consists of drug treatment checking, giving

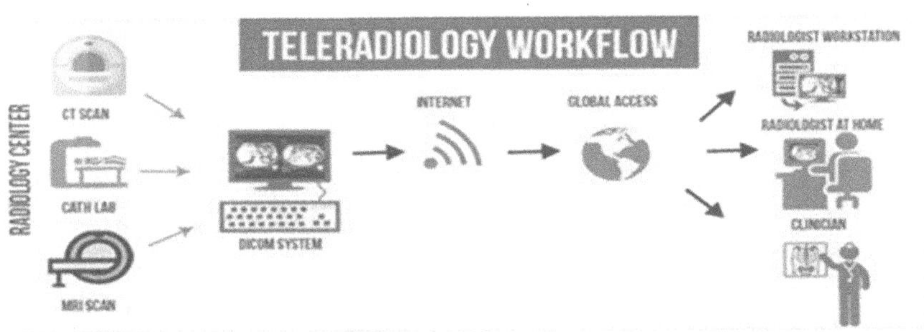

Figure 6.4 Teleradiology.

a prescription (Figure 6.5). Earlier approval and observing model consistency with the guide of remotely coordinating or video-conferencing. Furthermore, video-conferencing is immensely used in the drug store for different purposes, for example, giving the instruction, preparing the treatment, and playing out a few administration capacities [22].

A remarkable telepharmacy program in the US—directed at a government-approved local area, Community Health Associations (CHA), permitted the drug administering under central government programs. This program used video communication for apportioning drug and patient advising at six centers of different metropolitans and countries. There was one main drug store and five far off facilities in a few regions. The base drug center gives facility to provide drugs to the patients.

The remote site administering and understanding measure was portrayed as follows: when the data is sent from the far off facilities to the main drug store, the drug specialist confirms the printed copy and enters the request. The mark is likewise produced all the while, and the name line is sent to the far off-site. At the point when the name line shows up on the drug administering bureau known as ADDS, the approved individual can get to the medication from ADDS followed by prescription scanner tag

Figure 6.5 Telepharmacy.

examining, and the printing and filtering of names. When those means are done, the distant site faculty is associated with the drug specialist at the base drug store through videoconferencing for medicine confirmation and patient directing [23, 24].

6.7.2 Remote Monitoring

Distant checking of patients means experts screen a patient distantly by utilizing different innovative gadgets. This technique is essentially used for overseeing persistent health issues or explicit conditions of the patients, for example, coronary problems, diabetics, or asthma. These types of techniques used by administrations can give wellbeing results to customary in-person experiences, supply more trustworthy treatment to patients, and might be not costlier for the patients to get treatment as shown in Figure 6.6.

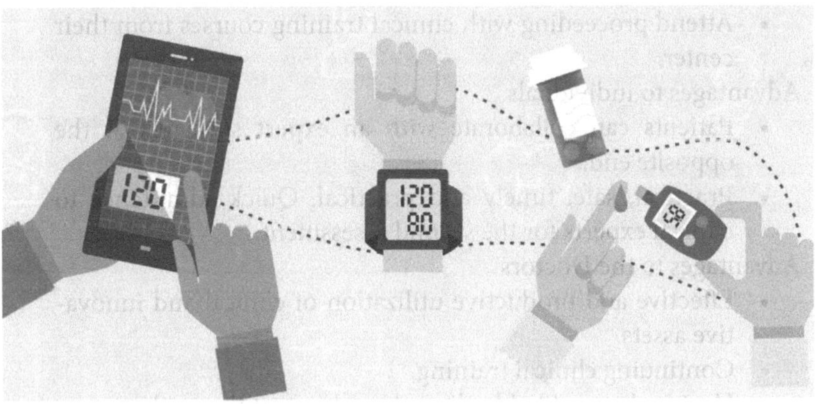

Figure 6.6 Remote monitoring.

6.7.3 Interactive Services

An inbuilt telemedicine administration gives continuous association among patient and experts. Like telephonic discussions, online appointments of the experts and home visits also will be provided as interactive services. Daily numerous exercises, for example, history audit, actual assessment, mental assessment can be led equivalently to those done in customer's eye to eye visits. Moreover, "clinician-intelligent" telemedicine administrations might be less exorbitant than face-to-face clinical visits [25].

6.8 Benefits of Telemedicine

Benefits the neighborhood specialist as he treats his patient without leaving his place

- Tele-radiology administrations will likewise be productive to the telemedicine focus.
- Video-conferencing likewise opens up additional opportunities for proceeding with clinical instruction or preparing for segregated or provincial wellbeing experts, who will be unable to leave a country practice to participate in expert gatherings or instructive freedoms.

To Rural Physicians and facilities (talked locales)

- Receive training from the subject matter expert/supplier.
- Better well-being results for their patients.
- Enhanced people group trust in nearby medical services.
- Attend proceeding with clinical training courses from their center.

Advantages to individuals

- Patients can collaborate with an expert specialist at the opposite end.
- Practical, safe, timely and practical. Quick admittance to clinical experts for the second assessment.

Advantages to the Doctors

- Effective and productive utilization of clinical and innovative assets.
- Continuing clinical training.
- Having the worldwide clinical local area at the work area.
- Enhanced Doctor-understanding relationship.

Advantages to the Patients

- Loved ones stay locally with family upholds.
- Immediate critical consideration.
- Confidentiality of forte assessment or visit (Because the patient visits the overall practice specialist, he can be seen for any strength care without any other individual knowing).
- Patient training courses (sustenance, oncology, and so forth)
- Properly balance out patient before transport.

Advantages to the Tertiary Care Institution

- Expansion of piece of the pie.
- Training for medical care experts, thus authoritative development of Services.

- Expansion of tertiary consideration organization.
- Providing support for all survey cases along these lines improving the standing [26].

6.9 Application of Telemedicine Using Machine Learning

There are a few mhealth applications that gather data about a patient's clinical history, way of life, and manifestations through a chatbot, and utilize this information to propose possible determinations [26, 27]. It is indistinct precisely how each demonstrative application applies AI methods as shown in Figure 6.7; however, a basic application would probably contain the accompanying fundamental advances:

1. Utilize a gigantic dataset of patients with their side effects and specialist educated or specialist checked analyses;
2. Prepare a model to foresee analyze;
3. Use the model to create expected judgments for the patient of interest.

6.10 Innovation Infrastructure of Telemedicine

Telemedicine is a method of transformation of Communication Technology, IT, Biomedical Engineering and Medical Sciences. The telemedicine system involves hardware, software and programming at both the patient and master expert end deal with a segment of the indicative supplies like CT-SCAN,

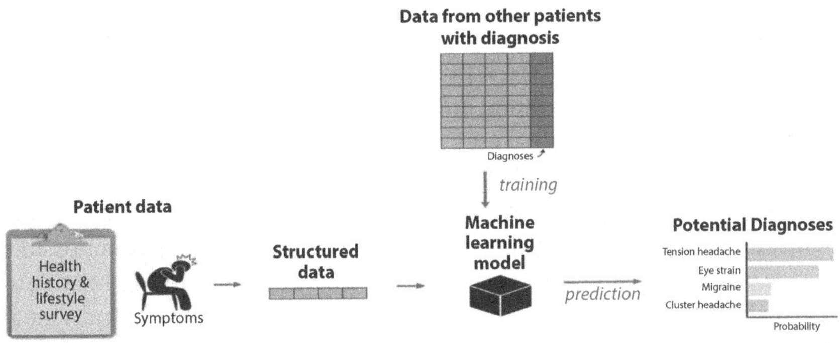

Figure 6.7 Telemedicine using machine learning.

ECG, X-bar and pathology Microscope/camera gave at the patient end. They are related through a Very Small Aperture Terminal (VSAT) system used with IT and forced by the Network Hub Station of ISRO [28]. Through a telemedicine system, the clinical pictures and other information identifying with the patients can be transported off the master subject matter experts, either early or reliably through the satellite association as Digital Data Packets. These packages are given at the master place, the photos and other information is used, so that the master expert can think about the data, perform an examination for treatment, contact the patient and propose the appropriate treatment during a Video Conference with the patient.

The telemedicine office appropriately engages the master trained professional, experts and the patient concealed by several kilometers to see by all accounts and speak with each other. This enables the master expert to study the physical and state of mind of the patient and suggest treatment [29, 30]. This far away teleconferencing and treatment are considerably more critical if there ought to be an event of post-action (Post Surgery) follow-up since the patient isn't expected to travel unnecessarily and from now on putting aside money and time. Consequently, the utilization of Information and Communication Technologies to the demonstration of clinical consideration rapidly and develops the exertion of the clinical arrangement.

6.11 Utilization of Mobile Wireless Devices in Telemedicine

Convention considered ZigBee in a framework that screens patients' physiological signs distantly. The framework was planned for use in PC helped actual recovery applications and walking observing. In contrast with Bluetooth, the ZigBee convention requires lower power utilization, bringing about gadgets that are less expensive to produce and lighter to wear.

Remote Local Area Network permits different processing gadgets in a restricted zone to impart among one another without physical cabling. The information transmission pace of remote LAN is better than PAN/BAN (for example 54 Mbps1 for IEEE 802.11g versus 3.0 Mbps for Bluetooth 2.0). In clinical settings, the transmission of complex information arrangements, for example, x-beam pictures or ECG in a remote LAN is possible (Figure 6.8). The framework permitted paramedic faculty to send patients' information (for example socioeconomics and ECG) utilizing individual information collaborators (PDA) and PCs from the patients' homes to

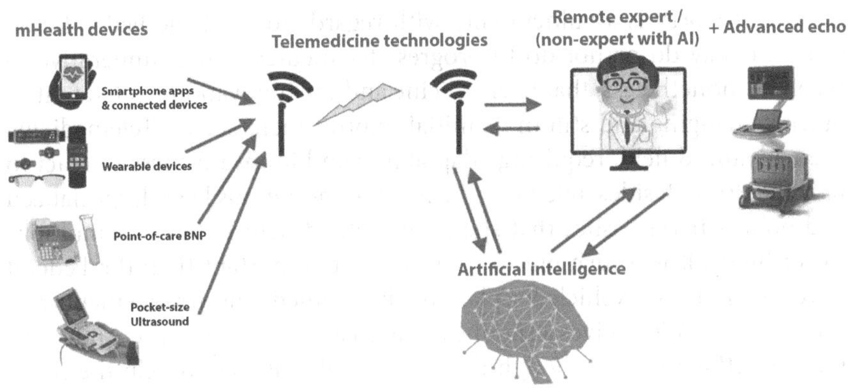

Figure 6.8 Workflow of MHealth.

ambulances stopped outside [31, 32]. Information was then re-sent to the trauma centers preceding patients' appearance to the emergency clinics.

The cell framework has advanced from absolutely a convenient phone framework, thinking back to the 1980s into an incorporated stage for portable voice and information administrations in the 21st century. Original (1G) was a simple framework that upheld just voice. It was supplanted by second-era (2G) guidelines including CDMA, TDMA, and GSM [33, 34]. Its primary distinction from the past age is the consideration of information administrations. For example, the Wireless Application Protocol (WAP) permits clients to get to site pages and applications from their cell phones and PDAs.

6.12 Conclusion

This chapter gives a depiction of the government's exercises in the concept of telemedicine. Telemedicine, in some structure, has been polished for more than thirty years. At the most higher level, a doctor is giving clinical treatment via phone is telemedicine. The main point discussed that telemedicine applications used various techniques picture, and sound abilities. These advances can go from the high goal actual pictures (e.g., x-beams) to modern intuitive video chatting frameworks. Although these high-level frameworks are quickly developing, huge doubt about telemedicine remains. This distrust is energized by worries about the current low utilization of telemedicine frameworks and about preceded with manageability once Federal, state, or private endowments vanish. At the core of a lot of this worry is a vulnerability with regards to what the Federal government may do or not do to progress telemedicine. It is imperative to recollect, nonetheless, that telemedicine and the innovation related to it, is quickly changing and still in its initial improvement stages. Telemedicine is an advanced field, requiring adaptability and innovativeness to react to its difficulties. Besides, telemedicine includes numerous lawful, specialized and policy-driven issues that should be settled before they can multiply. Accordingly, it will turn out to be even more important than the Federal government have a vehicle for planning its telemedicine arrangements and projects. The JWGT gives a significant discussion to achieve this undertaking. Nonetheless, even though the JWGT will embrace to help the savvy arrangement of telemedicine, Federal government organizations can't resolve all the issues examined in this report, alone. The different states, wellbeing experts and affiliations, and the private area should meet up to make telemedicine a suitable medical care conveyance choice for the United States.

References

1. Hailey, D., Jacobs, P., *Assessment of telehealth applications*, Alberta Heritage Foundation for Medical Research, Edmonton, Canada, 1997.
2. *Telemedicine Report to Congress*, North Dakota Law Review, *73*, 1, Article 7, 1997.
3. *Telemedicine Report to Congress*, U.S. Department of Health and Human Services, 13, 2000.
4. Afza, N. and Manjula, H.K., Opportunities and challenges of telemedicine system in healthcare sector—A conceptual study. *Int. J. Manag., IT Eng.*, *9*, 2, 128–140, 2019.
5. Mohandass, D. and Janet, J., An improved three pattern Huffman compression algorithm for medical images in telemedicine, in: *International Conference on Business Administration and Information Processing*, Springer, Berlin, Heidelberg, pp. 263–268, 2010, March.
6. Dwivedi, A.S., Mehta, D.S., Mehta, S., Dwivedi, S., Teletransmission of ECG Waveform: An Ingenious Low Priced Technique. *Indian Heart J.*, *36*, 24–32, 2012.
7. Kontaxakis, G., Visvikis, D., Ohl, R., Sachpazidis, I., Suarez, J.P., Selby, P., Rest, C.L., Santos, A., Ortega, F., Diaz, J., Pan, L., Integrated telemedicine applications and services for oncological positron emission tomography. *Oncol. Rep.*, *15*, 4, 1091–1100, 2006.
8. Angaran, D.M., Telemedicine and telepharmacy: Current status and future implications. *Am. J. Health-Syst. Pharm.*, *56*, 14, 1405–1426, 1999.

9. Clifton, G.D., Byer, H., Heaton, K., Haberman, D.J., Gill, H., Provision of pharmacy services to underserved populations via remote dispensing and two-way videoconferencing. *Am. J. Health-System Pharm.*, *60*, 24, 2577–2582, 2003.
10. Traynor, K., Navy takes telepharmacy worldwide. *Am. J. Health-Syst. Phar.*, *67*, 14, 1134–1136, 2010.
11. Conrath, D.W., Bloor, W.G., Dunn, E.V., Tranquada, B., A clinical evaluation of four alternative telemedicine systems. *Behav. Sci.*, *22*, 1, 12–21, 1977.
12. Muller, C., Marshall, C.L., Krasner, M., Cunningham, N., Wallerstein, E., Thomstad, B., Cost factors in urban telemedicine. *Med. Care*, *15*, 251–259, 1977.
13. Brecht, R.M., Gray, C.L., Peterson, C., Youngblood, B., The University of Texas Medical Branch—Texas Department of Criminal Justice telemedicine project: Findings from the first year of operation. *Telemed. J.*, *2*, 1, 25–35, 1996.
14. McCue, M.J., Mazmanian, P.E., Hampton, C., Marks, T.K., Fisher, E., Parpart, F., Krick, R.S., The case of Powhatan Correctional Center/Virginia Department of Corrections and Virginia Commonwealth University/ Medical College of Virginia. *Telemed. J.*, *3*, 1, 11–17, 1997.
15. Hayes, W.S., Tohme, W.G., Komo, D., Dai, H., Persad, S.G., Benavides, A., Juttner, H.U., Fleming, M.P., Wonsetler, B., Mun, S.K., Pahira, J.J., A telemedicine consultative service for the evaluation of patients with urolithiasis. *Urology*, *51*, 1, 39–43, 1998.
16. Harno, K.S.R., Telemedicine in managing demand for secondary-care services. *J. Telemed. Telecare*, *5*, 3, 189–192, 1999.
17. Harrison, R., Clayton, W., Wallace, P., Virtual outreach: a telemedicine pilot study using a cluster-randomized controlled design. *J. Telemed. Telecare*, *5*, 2, 126–130, 1999.
18. Bergmo, T.S., An economic analysis of teleconsultation in otorhinolaryngology. *J. Telemed. Telecare*, *3*, 4, 194–199, 1997.
19. Sparks, K.E., Shaw, D.K., Eddy, D., Hanigosky, P., Vantrese, J., Alternatives for cardiac rehabilitation patients unable to return to a hospital-based program. *Heart Lung: J. Crit. Care*, *22*, 4, 298–303, 1993.
20. Turnin, M.C., Bolzonella-Pene, C., Dumoulin, S., Cerf, I., Charpentier, G., Sandre-Banon, D., Valensi, P., Grenier, J.L., Cathelineau, G., Mattei, C., Multicenter evaluation of the Nutri-Expert Telematic System in diabetic patients. *Diabete Metab.*, *21*, 1, 26–33, 1995.
21. Albisser, A.M., Harris, R.I., Albisser, J.B., Sperlich, M., The impact of initiatives in education, self-management training, and computer-assisted self-care on outcomes in diabetes disease management. *Diabetes Tech. Ther.*, *3*, 4, 571–579, 2001.
22. Spencer, J.A., Dobson, D., Hoare, M., Molyneux, A.J., and Anslow, P.L., The use of a computerized image transfer system linking a regional neuroradiology centre to its district hospitals. *Clin. Radiol.*, *44*, 5, 342–344, 1991.

23. Eljamel, M.S. and Nixon, T., The use of a computer-based image link system to assist inter-hospital referrals. *Br. J. Neurosurg.*, *6*, 6, 559–562, 1992.
24. Rangasamy, M., Balasubramaniam, A.Y.Y.A.S.A.M.Y., Krishnarajan, D.E.V.A.R.A.J.A.N., Raviteja, A., Kante, N.R., Kumar, N.S., Role of telemedicine in healthcare system: A review. *Int. J. Recent Adv. Pharma. Res.*, *2*, 1–10, 2011.
25. Halvorsen, P.A. and Kristiansen, I.S., Radiology services for remote communities: Cost minimisation study of telemedicine. *Bmj*, *312*, 7042, 1333–1336, 1996.
26. Trippi, J.A., Lee, K.S., Kopp, G., Nelson, D.R., Yee, K.G., Cordell, W.H., Dobutamine stress tele-echocardiography for evaluation of emergency department patients with chest pain. *J. Am. Coll. Cardiol.*, *30*, 3, 627–632, 1997.
27. Giovas, P., Papadoyannis, D., Thomakos, D., Papazachos, G., Rallidis, M., Soulis, D., Stamatopoulos, C., Mavrogeni, S., Katsilambros, N., Transmission of electrocardiograms from a moving ambulance. *J. Telemed. Telecre*, *4*, 1_suppl, 5–7, 1998.
28. Doze, S., Simpson, J., Hailey, D., Jacobs, P., Evaluation of a telepsychiatry pilot project. *J. Telemed. Telecare*, *5*, 1, 38–46, 1999.
29. Trott, P., Blignault, I., Cost evaluation of a telepsychiatry service in northern Queensland. *J. Telemed. Telecare*, *4*, 1_suppl, 66–68, 1998.
30. Zelickson, B.D. and Homan, L., Teledermatology in the nursing home. *Arch. Dermatol.*, *133*, 2, 171–174, 1997.
31. Loane, M.A., Corbett, R., Bloomer, S.E., Eedy, D.J., Gore, H.E., Mathews, C., Steele, K., Wootton, R., Diagnostic accuracy and clinical management by realtime teledermatology. Results from the Northern Ireland arms of the UK Multicentre Teledermatology Trial. *J. Telemed. Telecare*, *4*, 95–100, 1998.
32. Parveen, N., Ali, A., Ali, A., Based Automatic Vehicle Accident Alert System, I.O.T., *2020 IEEE 5th International Conference on Computing Communication and Automation (ICCCA)*, IEEE, pp. 330–333, 2020, October.
33. Ahmad, A. and Ali, A., Performance Enhancement of Mobile Ad-Hoc Network Using Delay Aware Routing Protocol. *Int. J. Sci. Res. Sci., Eng. Technol. IJSRSET*, 1, 3, 187–192, 2015.
34. Kaur, R. and Ali, A., A Novel Blockchain Model for Securing IoT Based Data Transmission. *Int. J. Grid Distrib. Comput.*, 14, 1, 1045–1055, 2021.

7

Trusted Communication in the Healthcare Sector Using Blockchain

Balasamy K.

Department of Information Technology, Dr. Mahalingam College of Engineering and Technology, Pollachi, India

Abstract

The Internet of Things (IoT) is a booming research area in the recent world, which has many potential problems to solve. It grants anyone to interface anything with the web. Then again, secure association, correspondence and sharing of data are additionally under investigation in IoT condition. Communication and storage of information on any IoT device make genuine security issues. As the use of smart devices has steadily increased across different applications, this chapter concentrates on fusing IoT in the medicinal domain to upgrade the patient-focused clinical consideration. In any case, guaranteeing the security of the information produced and gathered by the IoT devices is the most testing issue in medicinal services administrations. The primary objective of this chapter is to design a framework for securing the IoT information produced in medicinal services. Thus, security mechanisms utilized in the IoT environment must be capable of making communications from end to end and must be embraced in low-cost IoT devices. The proposed framework achieves high throughput with less latency, which results in better efficiency.

***Keywords*:** Internet of Things, healthcare, security, encryption, medical data

7.1 Introduction

Healthcare has been an integral part of information technology (IT) as the industry has expanded considerably with the use of HER [1–10], remote

Email: balasamyk@gmail.com

Kavita Sharma, Yogita Gigras, Vishnu Sharma, D. Jude Hemanth and Ramesh Chandra Poonia (eds.) Internet of Healthcare Things: Machine Learning for Security and Privacy, (131–160) © 2022 Scrivener Publishing LLC

patient surveillance [11–24] and population health management instruments [25–32]. Because of increased cybercrime cases, the facts obtained from the above databases were lengthy and bulky, thus results in difficulty in the accuracy of medical knowledge, such as complex research, evaluation, and prediction, as well as a danger in data confidentiality [33–35]. Healthcare records have proved valuable to the health service. In addition, patients have the appropriate resources reported with their vantage point. Data security is, therefore, a key component of healthcare services and is a key element of data protection for sensitive data. Information concerning healthcare includes data of patients that cannot be submitted to any non-confident third party due to safety problems and misuse of the information [36]. This specific type of data consists of a list of patient data gathered in medical repositories from the outset of patient illness to recovery. Such data is also used in hospitals through several timed data. Medical data and medical facts, however, is distributed among different medical issues.

Inclusion with blockchain can make a major contribution to human health and well-being by the medical system [37–39]. Multiple patient and service provider signatures will be allowed on smart contracts providing access to or attachments of storage information in EHR is permissible only to registered users or appliances. This role guarantees that individuals can affirm the veracity of knowledge conveyed in EHR while also protecting the correctness of their true identities. Smart contracts provide for collaboration by mutual collages, as well.

In addition, smart contracts allow observers to have access to certain public health data as well as promote automated micropayments to help patients and healthcare providers active in a blockchain network. Nevertheless, the bulk of consumers who have been manufactured on the grid of the decentralized blockchain are constrained by sound technology and high development costs. The possibility of patient involvement by uploading their details to the publication of the public blockchain networks have also risen [40]. Blockchain designs is a distributed leading lumber to use an advanced process for encryption methods in the server (P2P platform) as shown in Figure 7.1. In 2008, this technology was suggested by Satoshi Nakamoto and was adopted the following year to shape the P2P electronic cash system, and these are the bitcoin cryptocurrency [40]. When introducing new decentralized technologies in the financial or non-financial areas [41, 42], the ethereum blockchain has developed the definition of a full programming language in the blockchain setting. Smart contracts are decentralized apps, which use and use a fully implemented language via ethereum or blockchain platforms [43].

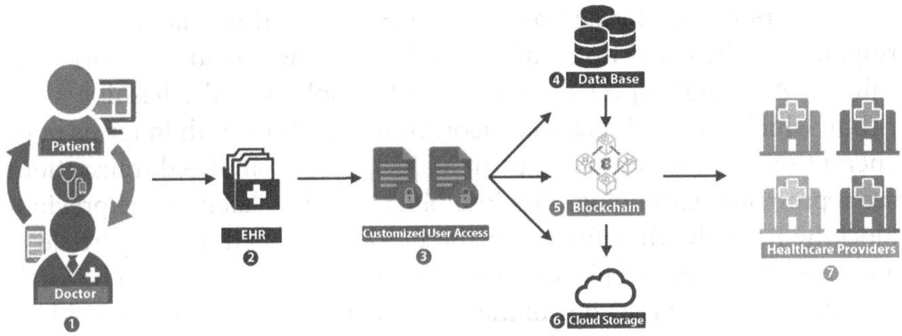

Figure 7.1 Blockchain infrastructure [18].

Issues with blockchain or smart contracts are about protection, anonymity, and scalability since the public is exposed to any transaction on the main network. As a result, there are many risks to data for this case. Although the program resides on a P2P network for smart contracts, the ethereum blockchain faces a significant problem in the implementation of particular use cases. Blockchain networks and smart contract influence are thus underperformed by the three parameters protection, privacy and scalability [44–47]. In the current systematic analysis, the goal is to provide useful insights into the technology, according to the researchers, to clarify the choices and vulnerabilities available in healthcare apps [48]. Subsequently, researchers will be able to plan a testing scenario by way of simple defining blockchain features that are embedded into healthcare products.

This chapter discusses the major differing implementations and blockchain networks. We address this convergence across different elements, such as modern architecture, device designs, process, scheme, paradigm, platform, approach, protocol, and algorithm, upon the mechanism designs of decentralized healthcare implementation. Since many of the issues at stake was the incorporation of blockchain technologies into healthcare, the solutions presented are being investigated to satisfy the demands of the system. This report also explores the merit, purpose, and challenges for emerging innovations, and offers suggestions for potential advancements and trends.

7.2 Overview of Blockchain

A blockchain is an encrypted block sequence that is validated by the network nodes and records a collection of transactions. In the block, events or transactions are checked by their timestamp (a time marker that corresponds to the exact time and dates the block was made), ensuring its validity. A central aspect of blockchain technology is the hashing algorithm, which is a unidirectional algorithmic function, both in terms of its encryption and its irreversibility. In a blockchain, completed transactions are represented chronologically in a block that is related to the previous block by its hash. Therefore, after the block is finished, it is a particle in an immutable chain of blocks referred to as the ledger, so it can be safely stored. Each user has a copy of the blockchain and can verify it quickly to ensure that it has not been compromised. Thanks to this encrypted, time-stamped, irreversible, immutable blockchain, it is possible to keep information safe and to guarantee anonymity regardless of how the technology is used [2]. For example, there is no need to meet, trust or know something about each other for two users who are carrying out a transaction together using a blockchain.

Blocks may safely provide personal information for dealing with the purposes of your data or control over the land, digital rights, or contracts. To have the blockchain for a new block, it must also be verified, authenticated, encrypted, which, for instance, involves the development of a new specific hash using the hashing algorithm for blocks that conceals all the details to safeguard it. Many that directly run the network or hack, as this technique is called, are called miners.

7.3 Medical IoT Concerns

Medical IoT-related issues can be categorized into three key categories, each being presented by the common public and that relates to issues in the security, privacy, and trust areas.

7.3.1 Security Concerns

These devices are vulnerable to different wireless/security threats because of the dependency of medical IoT devices around the practice in unsecured cordless communications. In fact, because of the lack of safety measures in which most clinical IoT devices, the perpetrator will eavesdrop and decrypt received data and information, either because of nature or due to inadequate security authentication mechanisms that could easily be prevented by a competent assailant. Another problem with protection is the potential to get unauthorized access even without detection, so such attacks cannot be known and stopped. This will result in higher rights, malicious codes being injected, or malware devices being infected. Medical IoT computers, on the other hand, maybe hacked and used to initiate attacks against Distributed Denial of Service (DDoS). In [7] how medical equipment is vulnerable to botnet attacks may lead to physical attacks on individuals. For instance, an attacker can potentially control a drug dose that kills a single patient or has serious health consequences. In addition, when seized by terrorists, medical IoT systems may be used as a form of targeted assassination. In addition, as stated in [7], medical IoT devices may have a detrimental impact on patient's psychological status, as patients may be frightened when they are rounded by robots instead of humans, which leads them to have a heart attack.

7.3.2 Privacy Concerns

Passive threats including certain traffic monitoring contribute to privacy concerns as information about the identities of patients may be gathered and revealed along with patient clinical confidential information. As an intruder can recognize his/her clinical history and health conditions, this is an extremely dangerous threat to patients that leads to dramatic serious impacts on patients. Another justification for compromising patients' privacy, by targeting hospitals, is identity fraud. Several of these actual attacks lead either to leaks or to the publication of confidential data compromising the confidentiality and privacy of the patients.

7.3.3 Trust Concerns

The violation of the privacy of patients transforms into a significant lack of identity. The idea of automation takes over the duties of physicians, nurses, and receptionists are becoming sceptical among patients. As a result, customers are more worried about having their health conditions monitored and controlled by a smart medical robot or a healthcare device.

7.4 Needs for Security in Medical IoT

IoT applications in real time gather vast quantities of personal data from company firms in hospitals, household, and financial transactions. To access user data, heterogeneous real time IoT presents a need for proper protection algorithms. While several security algorithms exist, they do not provide the IoT environment with higher-order security. It is difficult to perform complicated data encryption techniques in IoT devices due to the

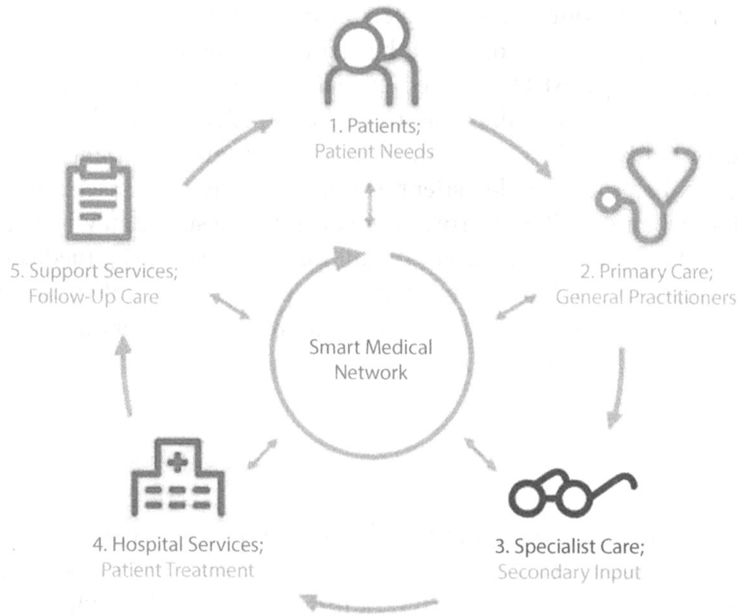

Figure 7.2 Smart medical network.

energy-limited nature of IoT devices. There is therefore a need for lightweight security algorithms that provide IoT data with a higher degree of security with minimal transmission and overhead computation on that kind of IoT devices.

IoT systems are a favored target for attackers because of the existence of applications, such as the storage of confidential data and connected to many network devices as shown in Figure 7.2. Additional vulnerabilities are emerging with IoT [6] beyond the real internet security problems. Specific applications can pose a serious life or business risk [15–17]. Imagine a scenario in which an intruder takes control of a car. Other threats include private information theft (privacy/confidentiality), information alteration (authenticity), and system malfunctioning (availability and resilience). Furthermore, an affected device can pose potential threats to entire secure connections, such as when a hacked device, along with a web camera, that is connected to the secure home network.

Unauthorized access to data and equipment raises concerns regarding patient safety in the healthcare system. Data confidentiality is one of the key aspects of protection that must be taken into consideration when sharing data. In addition, it is important to minimize the high computation overhead caused by existing cryptographic security algorithms and to withstand different attacks such as Brute Force, Eavesdropping, Man-in-the-Middle, etc. This creates a need for reliable, effective, and lightweight security algorithms to be created. It is therefore important to build lightweight algorithms to monitor and manage data protection across different layers of the IoT architecture.

7.5 Uses of Blockchain in Healthcare

In many industrial fields, the functions of blockchain have developed different advantages and can be an effective tool for implementations in healthcare systems. However, unless the application needs decentralization, blockchain technology could not be the perfect option to solve any problem in industrial fields. Blockchain applications may be helpful if an

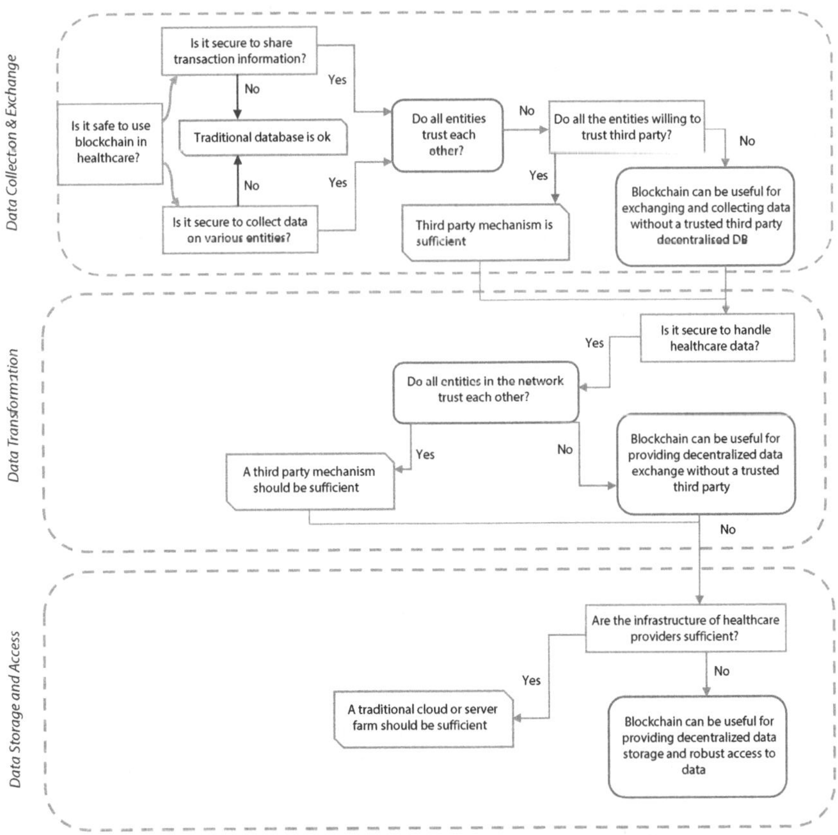

Figure 7.3 Blockchain usage in the healthcare sector.

application is it needs to be decentralized. For example, when an application is improved by running on a P2P network of computers instead of an individual system, this circumstance arises. While it might not be necessary to decentralize some healthcare applications, most of such applications may have profited from a centralized, untrusted one. However, such traits should be explained before determining if a blockchain may be included in the application of medicine [49–58]. A general scheme of evaluating the use of the blockchain in the healthcare sector was discussed as part of blockchain innovation is shown in Figure 7.3.

First, it is required to convert knowledge into trustworthy environments. Present data transformation methods may be used to execute different methodological activities automatically, speeding transactions between sides [59]. While a conventional network can be created, two drawbacks have appeared in healthcare application tools. These factors also include higher trade fees than public blockchain products, and those fields can be nearly entirely misinterpreted by protection, internal policies, and ethics [60].

Secondly, the confidentiality and tamper-proof properties of blockchains can help logs regarding public transactions. It contains records of timestamps that all parties who communicate with the blockchain will publicly report and inspect. These logs. Health programs should strictly follow the notion and process them in a blockchain network, each transaction engages in proper review and follow-up. One value relevant to health application is the availability of P2P connectivity to shared data between industrial process involving parties. In the EHR systems, this condition is extremely normal for each other to find out about the practices or duties.

Finally, the efficiency of the robustness of the distributed system can be attained. Potential alternatives may meet the needs of applications run by clouds and equipment networks. However, several issues stem from confusion in the company which runs its systems, such as the cloud, or client privacy requirements [61]. Trust agencies need to be credited for the facilities providers.

7.6 Solutions for IoT Healthcare Cyber-Security

Although the development of IoT architecture is seen as the model of fog/edge computing is considered as the protection solutions research is still at the beginning era [9]. This chapter, therefore, concentrates on delivering new approaches aimed at enhancing safety in medical IoT devices that are light enough for coexistence and complementary at a reasonable overhead and footprint cost in the same facilities. Such methods are as follows:

7.6.1 Architecture of the Smart Healthcare System

The amount of IoT healthcare-connected medical users and devices are on the rise. There is a significant development in the relation of the room of patients, including medical equipment, room lighting, air conditioners, beds of patients, etc. The healthcare IoT system's physical and cyber world interconnections are in four distinct layers as shown in Figure 7.4.

(i) Data Perception layer
(ii) Data Communication layer
(iii) Data Storage layer and
(iv) Data Application layer.

7.6.1.1 Data Perception Layer

The main task of this layer is to deal with personal clinical information that is collect data from the user in real time. Medical information are collected through various short-range communication technologies such as Radio Frequency (RF) Waves, Z-wave, Near Field Communication (NFC), ZigBee, Bluetooth etc., and transfer to the upper layer for processing these data. Sensors such as RFID tags are incorporated into the Medical IoT devices and contain all the information that can be scanned to receive all the related data. Once the scanned data is transferred to the upper layer in the hierarchy, i.e., the networking layer, the data is processed accordingly.

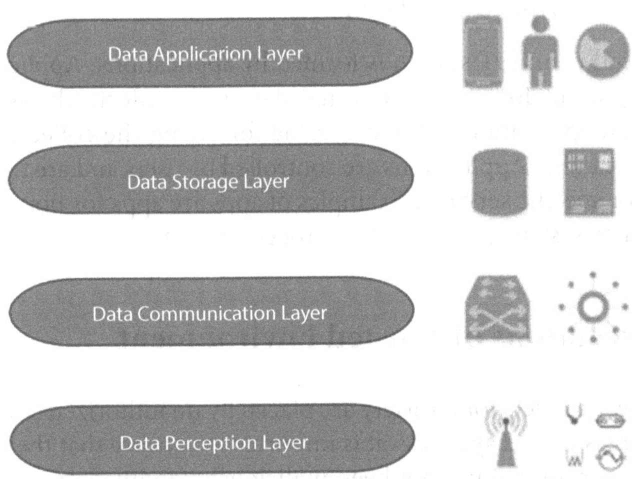

Figure 7.4 Layers of healthcare security.

Medical devices such as pulse oximeters, surgical lights, blood pressure monitors, glucose monitors and other devices in the patient room can be connected to the IoT gateway using the above-mentioned short-range communication technologies. The medical information gathered in the perception layer is encrypted using different techniques.

7.6.1.2 Data Communication Layer

It is the main channel between the other layers in IoT architecture and the different IoT operating activities. The whole physical structure has a lot of patient evidence and knowledge to communicate with other nodes. It is therefore necessary to establish an appropriate network of connections between these nodes through a communication protocol. Based on the designer's protocol, the interaction could be wired or wireless. In addition, networks are extremely critical IoT elements that bind the external world of the internet. Intelligent network infrastructure is required for IoT.

7.6.1.3 Data Storage Layer

The patient data contained in the Cloud Medical Server (CMS) is managed in the storage layer. Remote user authentication for accessing IoT data is handled by the Cloud Authentication Server (CAS) and the Central Cloud Server (CCS) stores the other metadata. The obligations of this layer are also the identification and management of IoT user data.

7.6.1.4 Data Application Layer

Another end of an IoT system is formed by applications. Applications are vitally important for all the collected data to be effectively used. These cloud-based applications are responsible for giving the collected data an effective meaning. Applications are controlled by users and are the point of delivery for specific services. Examples of apps are apps for home automation, security systems, industrial control centres, etc.

7.7 Executions of Trusted Environment

Since many products for a variety are placed by unauthorized personnel in outdoor areas with easy access, it is important to ensure that the executing software environment has not been maliciously modified or transformed. A root-of-trust (RoT) environment and, finally, a chain of trust (CoT) with authentication and encryption [6] can be generated using strong authentication system protocols and its use of dedicated hardware and cryptographic techniques.

7.7.1 Root of Trust Security Services

Global Platform has identified nine services that could be provided by Roots of Trust, based on its experience with the devices and markets addressed by Global Platform through its members: authentication, confidentiality, identification, integrity, measurement, authorization, reporting, updating, and verification.

With one or more security services, one can create a Root of Trust. However, unless it reveals the interface to the service in any way, the Source of Confidence cannot claim to deliver the service. For instance, consider the deployment of many Protection Services by a Root of Trust where the Verification Service can preserve the public key shielded position and be able to check the legitimacy of the next item in the Chain of Trust without revealing the Integrity Service and Measurement Service's protected capability. In this scenario, the RoT is just verification RoT and thus cannot pretend to be a secrecy or calculation RoT, because such facilities are exclusively customized to and reserved for using the verification RoT.

To securely store and protect sensitive information, a Root of Trust can retain one or more protected areas used. A protected area should be assisted by an interface that offers an acceptable degree of security to control access to sensitive information and discourage unwanted dissemination, usage, and use (for example, sensitive data stored in a protected area should only be accessed by an API when the API verifies the right of access to this sensitive information in an implementation). Certification policies, or the willingness of a vendor to provide value-added features for their goods, can determine the level of physical and logical security of the shielded site.

These resources can be used in the first boot code for safety-related operations, for instance, to check the credibility of others before the control is turned over, or to document potential comparison and monitoring network characteristics. Each part in the sequence can easily implement additional RoT Protection Services and use them to conduct security-related tasks, such as approving access to protected resources, upgrading essential system firmware, or reporting platform functionality, as the device boots it's the way through initialization. Certain protection services may use RoT Security Services to secure the keys, information stored on computers, and data in transit (e.g., Trusted Service Manager) to and from back-end services. Developers of applications can either specifically or indirectly use RoT Security Services via the use of these other security services.

For protected storing and maintains the privacy of at least one certificate, the RoT for authentication retains one or more shielded locations. One or more access controls are stored here by a service user for later use to help authenticate compliance with a protocol, such as performing access control over the use and/or release of keys or authorizing access to data. An app is also included in the Root of Trust for Authentication to preserve permitted access to and use of shielded location information, as well as protect it from unauthorized use and disclosure. Registered usage, in this case, is for authentication purposes only.

A RoT factor is necessary for the stable boot phase to conduct code verification in a trustworthy manner at the boot stage of the first unit. Figure 7.5 explains the phases of general creation and application testing. First, as with any development environment, the developer compiles the source code that produces the binary image or firmware of a computer. Second, the binary image's hash digest (a number code that can identify it uniquely) is determined. Thirdly, the public key pair that comes from a private key, encrypts the hash digest. The effect is a signature that, along with the firmware, must be put on the computer. The developer must also write a public key to the computer's writing-once memory. The bootstrap

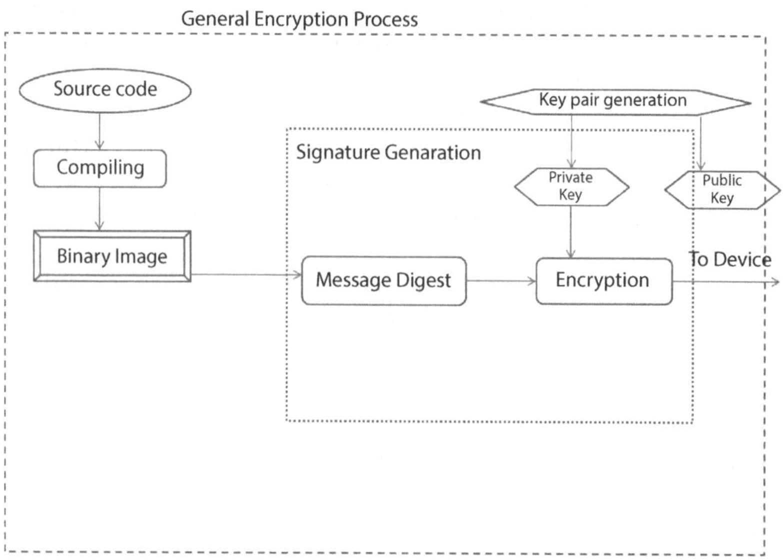

Figure 7.5 Root of Trust.

program performs on-storage hash digest recalculation firmware during initialization. The final stage is to use the public key to decode the signature and equate it to the firmware digest.

7.7.2 Chain of Trust Security Services

There are two kinds of confidence relationships in the Chain of Trust: Internal Chain of Trust and the External Chain of Trust. An Internal Chain of Trust is instantiated by a platform as it executes a set of code modules in an environment in which the effect of any inspection of code or data integrity that might take place is difficult to log. On the other hand, as it intentionally extends a security service in such a manner that you can check the links in the chain generated by the extensions at any time and validate their validity, a platform instantiates a Clear Chain of Confidence, thus verifying the legitimacy of the security service it extends from end to end. The External Chain of Confidence exists through a set of code modules in which a RoT tests the next code module for authentication and authorization. If it passes such tests, it can be executed and becomes an eRoT in an external Chain of Trust.

Currently, an External Chain of Trust expands service to other modules from a RoT (either a Main or a Secondary RoT) or module. The Chain of Trust for Authentication, under which the RoT requires a verification

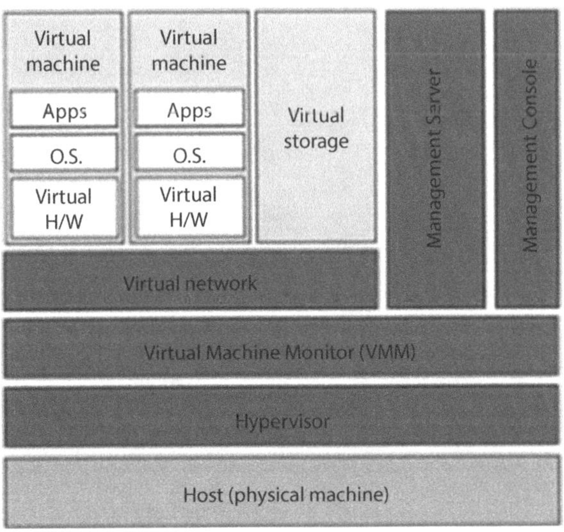

Figure 7.6 Chain of Trust.

service, is a clear example of this, and it uses that service to validate a module providing a verification service close to that of the parent verification service. In a RoT, an explicit chain of confidence will always start. Explicit Chains of Trust, including Roots of Trust, must contain a minimum of one security service. The approach of expanding the Root of Trust service to subsequent modules is strongly dependent on implementation.

This method is a multi-layered framework, which includes a hypervisor. Next, the boot loader is validated by the hardware. If efficient, it is treated as a trustworthy part and the next boot step can be checked. In this case, the next move to trust is the hypervisor. The hypervisor begins and tests its fields until they are booted until confirmed as shown in Figure 7.6. The coexistence of trustworthy and unverified realms, using the hypervisor to implement the distinction, is feasible in this case.

7.8 Patient Registration Using Medical IoT Devices

The architecture of the smart healthcare system is represented in Figure 7.7, which contains security phases starting from patient registration till he leaves the hospital. The process for registration is a one-time process. It is running after the installation of the medical devices and devices for users. Three types of registration processes exist. Since various methods and inputs are used for medical and user device registration, registrations

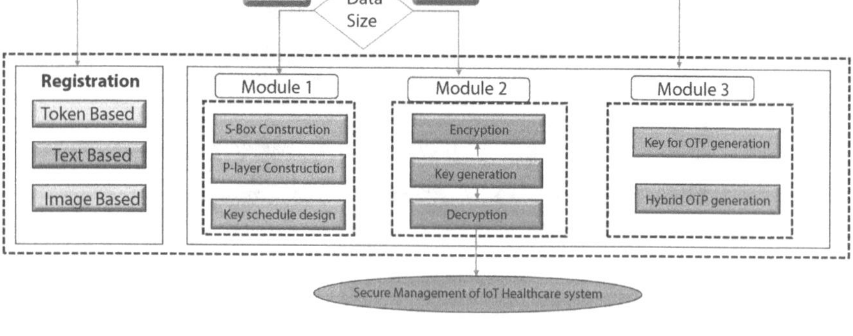

Figure 7.7 Architecture Diagram of Smart Healthcare.

based on text, token and image are performed. The parameters used in the registration process are Patient ID, Password and Device Product Code.

Patient ID (PID): For every person, it is a proposed unique identification number. It is similar to the Aadhaar number in India and Social Security Number (SSN) in the USA. Every patient should have a Unique ID that is registered in the Medical IoT environment.
Passcode (PC): A passcode, also referred to as a password, is a memorized secret, normally a character string commonly used to validate a user's identity. The patient and user can select the password based on their circumstance and wish, with the conditions specified in the portal.
Device Product Code (DPC): This is a number of 96 bits, built as a global identifier. It gives any physical object a unique identity. It is a format that is open and can be encoded on RFID tags. Tracking all sorts of objects, including fixed properties, records, business goods or recycled transport items, can be done using device product code.
Text Based Registration: During the induction process, every patient outside the hospital sector must be registered. Of the patients, the PID and a PC are collected. A hash function to produce a hidden Patient Unique Number (PUN) is added to them. These data are preserved for further reference in the archive of the cloud application. After a patient's satisfactory registration, an acknowledgement will be submitted to the patient stating that the registration has been done successfully.

7.8.1 Encryption

Encryption is a process that encodes a text or file such that certain users can read it only. Encryption means encrypting, using data and a key to convert it into the secret message and at the receiver side decrypt, or decode, the received data. The message is referred to as plaintext in a hidden text. In its encrypted, unreadable form, it is called ciphertext. A sequence of numbers used to encrypt and decrypt data is an encryption key. Algorithms are used to construct encryption keys. It is random and special to each key.

There are two types of encryption systems: symmetric encryption and asymmetric encryption. Symmetric encryption is used to encrypt and decrypt data with a single password. For encryption and decryption, asymmetric encryption uses two keys. A public key, which is exchanged between users, encrypts the secret. Data is decrypted by a private key, which is not exchanged.

7.8.2 Key Generation

Using symmetrical key cryptography, key control triggered nightmare for the parties. They were concerned about how to relay the keys safely and reliably to all users so that the message could be decrypted, which offered the opportunity to intercept the keys in transit for third parties to decipher the top-secret communications. Thus, the entire coding scheme was corrupted if the key was compromised, and a "Secret" would no longer remain a "Secret." So there exists a concept called key generation, where the required key for encryption is generated using random number permutation.

7.8.3 Security by Isolation

Security by isolation is a conventional security strategy that has proven to be effective over time [8]. It may be applied using multiple physical types of equipment or technology, implementing policies of usage, to reduce the supply and access to services. The separation of separate physical devices involves duplication of energy, which means higher costs. Maintaining two machine working models in separate processors and memory, for example. Virtualization could achieve logic isolation and can simulate a network to access the underlying domains or application hardware while preventing replication of resources. In addition, any form of connection, such as data dependencies or synchronization, may be necessary for the functionality of a given programme, meaning the entities concerned cannot be entirely segregated. Many IoT implementations would involve coordination with sensors and network transmission, making it less desirable to complete isolated systems.

7.8.4 Virtualization

The ability to separate programs (code and data) to share similar memory is spatial isolation. A feasible approach to minimize costs is to make it easy during the deployment of applications in memory sharing. Through the address bus, data bus and control buses, the memory is linked to a processor and applications that share the common physical address space. It is thus possible to enforce correspondence through shared memory than transferring messages. A running program, however, may have access to the address space of all processors without sufficient isolation, i.e., a malicious or malfunctioning code may enter any memory location. For memory provision, the processor must be fitted with a hardware block separating it, which provides the device with virtual memory abstractions, called a Memory Management Unit (MMU). Initially, in both processors (desktop and server), the MMU was available, being essential for the installation of complex and reliable operating systems.

The generated memory addresses are referred to as virtual in these processors addresses (VAs) and physical addresses (PAs) need to be translated by the MMU, handled by the host OS or virtualization layer. The OS holds a table to operate, called a page table, which maps VAs to PAs. For each virtual address, the MMU makes interpretations with a Translation Lookaside Buffer (TLB), transforming VAs to PAs. The most regular translations are preserved by TLB entries, preventing memory access from the page table intrusion. A trap is provided to the OS when there is no efficient translation (the event of a page fault). Mischievous or flopping memory access is thereby established, and it is possible to terminate the liable operation.

The distinction between VMs or domains must be maintained while the virtualization layer is present, while the guest OS (virtualized OS operating on a VM) should be capable of preserving the separation between processes. By establishing a new level of abstraction in the interpretation of the memory location, this is possible. VAs is converted to hypervisor-managed Intermediate Physical Addresses (IPAs). However, additional MMU hardware features may be required for the efficient management of IPAs. The IPA management must be emulated by the hypervisor without sufficient hardware assistance, through the shadow paging method, which preserves the IPA and PA relationship in an in-between page. Without proper support, virtualization on processors gives the resulting hypervisor extra complexity, and often prohibitive overhead.

The need to maintain an equal time allocation between programs based on the execution priorities of the processor is temporal parting. Temporal parting is achieved by software, unlike spatial separation, which does not need any specific hardware configurations, regardless of the addition of functionalities by some processors to boost the system performance. This technique originated as a method for enhanced processor use with the first OSs and to permit several programs and operators for resource sharing of the device concurrently. For two separate factors, in ES, temporal differentiation is a problem. First, many ES have extreme time limitations, and this must be properly resolved by temporal separation. Second, by hogging the CPU, a mischievous or flopping program may decline or even hinder the other software execution.

A program block named the scheduler, which is deployed for both the OSs kernel and the hypervisor is used to enforce temporal separation for virtualized systems. Hypervisors create a virtual CPU (VCPU) abstraction, which is connected to virtual machines. The Virtual CPU is a physical processor machine illustration that permits multiplexing the time. The scheduler then executes an algorithm that must pick the next one to run from the available VCPUs.

Operating networks have examined and manufactured virtual equipment developers, programmers of languages, developers of compilers and hardware. While any virtual machine implementation has its specific characteristics, there are also fundamental principles and technology like all virtual machines. Since the numerous systems and technology of the virtual machines have been created by different parties, this set of information and the core technologies of the various types of virtual machinery are particularly important to unify. The fidelity whereby a structural interface is applied by a virtual machine is also connected to architectural boundaries and is a major factor in designing a virtual machine. Consequently, device architecture is useful to describe and to sum up. A whole system is supported from the viewpoint of the operating system by the machine underneath. A method is a total execution system that can endorse a variety of potentially important processes at the same time various users. A file system and other I/O resources are common to all processes. With the processes coming and going, the system environment persists over time (with reboots occasionally). The framework allows processes to communicate with their resources with an OS that is part of the system and assigns physical memory and I/O resources. Therefore, the computer is only controlled by the underlying hardware from the system's viewpoint, as well as the ISA offers the communication between the device and the system.

Hypervisor based system isolation methods have the benefit of greater risk security, such as root phones, since a user who roots his OS device instance does not jeopardize the corporate OS instance. However, these hypervisor methods usually have better efficiency or battery impacts and can often be more cost-effective since a corporate OS picture has to be created and secured. Container approaches often benefit from seamless user experience because certain aspects of personal and business users can interact in ways that are never possible for separate OS instances. The compromise is usually the need to manage the underlying OS instance to prevent root attacks on the device from protecting the application layer container.

There are three major hypervisor deployment approaches are adopted: (1) Full virtualization (FV), (2) para-virtualization, and (3) virtualization hardware supported. This technique may be used independently or by integrating two or more techniques with a composite approach.

Full Virtualization includes a processor that facilitates the installation of a hypervisor capable of offering virtualization on the guest OS without any alteration. This helps unmodified visitors to virtualize and carry out more tech activities. But the hypervisor must obey preferential instructions which can lead to performance penalties since the time taken for emulating trap mechanism instructions are thousands of times longer than executing the intrinsic instructions.

Different methods that do not include processor architectures in hypervisors should not match the specified ISA. A workaround is to change and execute the guest OS, resulting in para-virtualization, before its launch in a VM. These modifications replace hypercalls for critical instructions (instructions that alter the actions of the processor) or other OS modules, such as the I/O subsystem. Hypercalls are utilities that the hypervisor provides that are invoked by the guest OS. They are similar in OS to machine calls (syscalls). In addition, hypervisor functionality can be expanded by using PV, for maximizing the inter-VM communication implementation. Additionally, PV is also used to boost performance in environments where performance problems are faced by completely virtualized hypervisors.

In history, the ARM architecture is not virtualizable, because when executed in unprivileged mode, some sensitive instructions do not trap. However, as an ARMv7 architectural expansion, the most current 32-bit ARM processors, including the Cortex-A15, have virtualization hardware support. There have been a variety of research ventures aiming to while they need different degrees of para-virtualization and have not been stabilized, they allow virtualization on ARM processors without hardware virtualization support. KVM/ARM is primarily designed to work with the virtualization extensions allowed to run the unmodified guest OS on ARM processors. The need for stronger and easier hypervisors led to hardware support for processor vendors, which culminated in hardware-assisted virtualization. A variety of methods may be used to support hardware. Various rights permitting hypervisor, application, and operating system kernel with multiple authorizations; and intrude virtualization for enhanced device hypervisor power. Therefore, the vendor of the processor decides the hardware-assisted support standard, the functionality required and the costs depending on the market niche.

7.9 Trusted Communications Using Blockchain

The center of IoT creativity and, at the same time, the biggest danger of its acceptance is communication skills. The blockchain [10] is a promising technology that aims of supporting the integrity of data and reliable communications. As a level of protection, it is focused upon decentralization.

To carry out verifiable, safe and it operates by keeping a list of documents linked by digital signature via time-stamped collaboration with external parties. This technology has some appealing features that make it suitable, despite all threats and vulnerabilities involved in the IoT, for potential developments.

Blockchain is an IT mechanism that requires a community of untrustworthy actors to validate transactions. It offers a distributed ledger that is immutable, open, secure and auditable [1]. This is essentially a distributed archive of all transactions or digital activities carried out and exchanged by the parties concerned. When information is entered into the blockchain, it can never be changed or erased. A certain and verifiable record of any single transaction ever made is included in the blockchain [2]. On a blockchain network, a new block is automatically formed when the next transaction is formed. A new block is immediately connected to the previous block, the original block, using Hash in the previous block. Once a new block is created, they are sent to all connected computers involved in the blockchain network in real time.

7.9.1 Record Creation Using IoT Gateways

Blockchain multi-layer can merge the features of blockchain and IoT, providing solutions to all issues which cannot be addressed by using only one technology. By integrating these two innovations, blockchain offers a next-generation IoT platform based on blockchain technology. The key benefits of this architecture are the solving of the existing blockchains, particularly in the absence of scalability. The IoT devices may be the private blockchain's nodes, some of them even belong to the next existing blockchain layer. The whole network will be involved in this case. The sensors and their gateways are now the nodes in the private BC (Blockchain); therefore the miner is the gateway since it is the most important node in its private blockchain. Before transmitting data using public and private keys, each system needs to authenticate with the network. Both keys are unique to each one. The gateway stores all keys in the local storage so that any system authenticating with it is quickly recognized. There may be local storage for each private blockchain. Notice that only filtered data that represent essential conditions can be stored in the cloud, processed and irregular data. The computer begins constructing a new block after the registration is complete. The patient's private BC will be attached to this block until checked by the gateway. All captured data is stored in the blockchain (local storage portal) database. The portal processes the medical data and regularly generates diagnostic knowledge.

7.9.2 Accessibility to Patient Medical History

The documents containing patient data have already been deposited in the patient's cloud service in this case. A record that has already been preserved must be accessed by the gateway/authorities that have authorisation. The core work, using a public blockchain, is concentrated in layer 3 where all cloud services are linked. Access and authorization management is achieved using the cloud contract. By using his account, the patient has full access to the data in the cloud; he is simply the owner of the data. The cloud gives back the ID of this record as it holds the record. The patient can access this record through this ID. The cloud then produces a transaction to remember that on this day, the patient has accessed the record of that ID. The authority sends its ID to the cloud that is acceptable and waits for the ACK. In this case, the ACK is the result of learning that the ID of the authority is one of the authorized IDs in the contract of the cloud list.

7.9.3 Patient Enquiry With the Hospital Authority

When a patient visits an agency, a new block is added to the second level of blockchain. As a result, both officials will add a certain block and a similar block will also be added to the blockchain supplier of the cloud. When the patient ends his visit, the authority attaches to the public blockchain, a new block that contains the authority's ID, the patient's ID and any information about the data contained in the authority's blockchain storage. The visiting authority creates a block in the blockchain of the cloud to remember that the patient of this ID has visited the proper authority. It also mentions the location where the data was stored.

7.9.4 Blockchain-Based IoT System Architecture

The architecture is composed of three layers as shown in Figure 7.8.

7.9.4.1 *First Layer*

This is the "patient" layer; it contains all the IoT nodes that obtain information about a single patient; medical information or some other form of information that explains his environment. A strong device that will serve as a key to the higher layer blockchain is a primary node in this blockchain. Notice that there would be a proper blockchain for each patient.

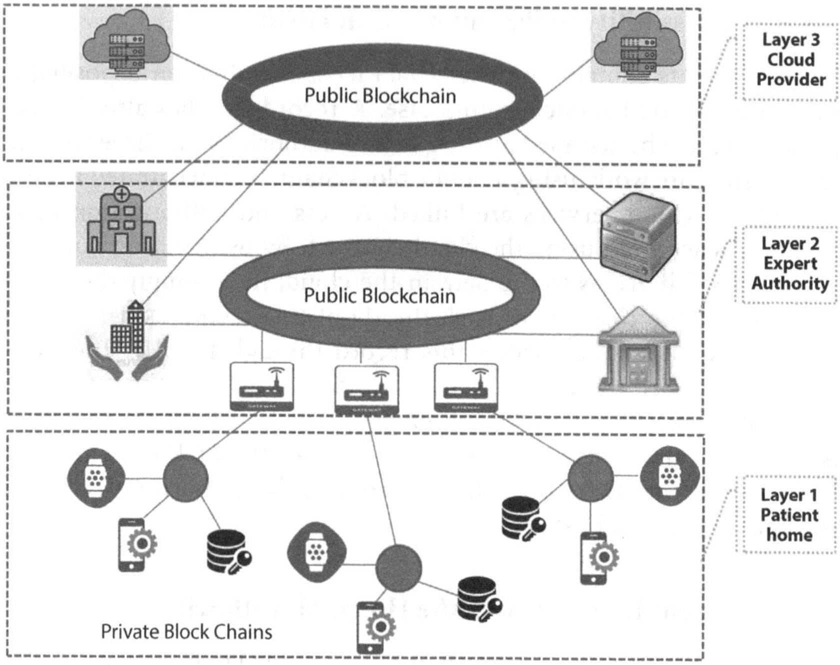

Figure 7.8 Blockchain based IoT smart healthcare.

7.9.4.2 Second Layer

This is the central layer or "authority" layer; it includes representative nodes for all medical stakeholders that are involved in patient-related data, such as hospitals, medical facilities, laboratories, etc. Members of this blockchain are also the gateways in the first layer.

7.9.4.3 Third Layer

This is the upper layer or the foundation of the cloud supplier. IoT machines are typically completed by cloud computing and storage capabilities. A domain-level blockchain is important to monitor the relationship between cloud services to allow each other access to patient data wherever it is stored.

7.10 Combined Workflows

7.10.1 Layer 1: The Gateway Collects IoT Data and Generates a New Record

The whole network will be involved in this case. The sensors and their gateways are the nodes in the private Blockchain, so the miner is the gateway since it is the most powerful node in its private blockchain. Before you begin sending data via public and private keys, each device must authenticate with the network. Both keys for each system are specific. The gateway stores all the keys in its local storage, so that any computer with it will quickly be identified. There may be local storage for any private blockchain. Note that only screened, processed and irregular cloud-based data can be stored.

The computer begins to build a new block after completing the registration. This block is attached to the patient's private BC until the gateway is validated. All collected data is kept in the database of gateway local storage. The portal processes the data and regularly generates diagnostic data. The gateway must however be registered remaining blockchains with two layers: layer 2 for correspondence with various types of authority, and layer 3 for the saving of periodical documents and emergency data. In the next phase, all details affect just the normal/emergency documents.

We primarily communicate between patients and various forms of bodies in layer 2 blockchain, and interaction can also occur between bodies themselves. Here we have job facts and stake proof to verify any transaction based on the previous transactions. This implies that many miners could mine and validating the blocks. The Gateway establishes a new public block in the BC after registration (Ethereum is a safe choice) to tell the healthcare specialists responsible for the treatment. In a cloud layer, the Gateway stores the record.

7.10.2 Layer 2: Gateway/Authority Want to Access Patient's Medical Record

The patient's cloud service has already saved documents that include patient details in this case. The approved gateway/authorities must access a record that has been already saved. The principal activity is focused on level 3 if connected by public blockchain to all cloud providers. Access and authorization monitoring is carried out via the cloud agreement. By using their database, the patient has full access to the cloud data; it is the data holders. The cloud returns the ID of this record as it saves the record. The patient can view the record by using this ID. The cloud also produces a transaction that states that on that date the patient has accessed that ID. The Authority will give its ID and wait for ACK to the right cloud. The ACK is the result in this case of the discovery that the ID of the authority is one of the authorized IDs of the list in the cloud.

7.10.3 Layer 3: Patient Visits and Interact With an Authority

When a patient meets an authority, a new block is added to the layer 2 blockchain. Consequently, both officials will obtain the same block and the cloud vendor blockchain will be substituted by a similar block. If the patient has finished his visit, the authority attaches to the public blockchain new blocks containing the identification of the authority, the patient identification number and some details contained in file storage. In the cloud block, the visited authority generates a block when the responsible authority visited the patient of this ID. It also notices the location of the storage data.

7.11 Conclusions

The purpose of this research is to discover new IoT-driven healthcare technologies, sensors, or systems to provide healthcare practitioners with improved knowledge, helping them to take action to prevent and diagnose diseases. In addition, our research opens a forum for data protection experts to do further research to mitigate IoT healthcare sensor and system vulnerabilities. Healthcare is a major industry in the country, and since the IoT revolution, healthcare systems have become more effective and secure. Health experts and scientists have improved the diagnosis and prevention of diseases with the exponential development of IoT products. Therefore, multiple sensors and systems powered by IoT have been launched and some are now being used successfully in hospitals and for personal healthcare. In a relatively short period, these instruments gather a vast volume of data. Therefore, during the use of such instruments, data breaches may occur.

We found that the most important advantage of using smart healthcare technologies and systems was performance. Around the same time, we have found that the protection and privacy of personal data were the most critical problem when using IoT healthcare devices. Researchers interested in biomedical science may focus their studies and experiments on advanced IoT sensors to detect health problems and diseases by measuring various body parameters. To ensure the secrecy and credibility of the information, information management professionals and IoT-driven healthcare manufacturers are advised to concentrate further on establishing users' protection and privacy. Future researchers may pay further attention to the development of IoT-based devices or sensors that help other body systems-related medical conditions that have not been described in this segment.

References

1. Balandina, E., Balandin, S., Koucheryavy, Y., Mouromtsev, D., IoT use cases in healthcare and tourism, in: *Business Informatics, CBI, 2015 IEEE 17th Conference on*, vol. 2, pp. 37–44, IEEE, Portugal, 2015.

2. Pang, Z., Yang, G., Khedri, R., Zhang, Y.-T., Introduction to the special section: Convergence of automation technology, biomedical engineering, and health informatics toward the healthcare 4.0. *IEEE Rev., Biomed. Eng.*, 11, 249–259, 2018.
3. Bogaz Zarpelão, B., Sanches Miani, R., Kawakani, C.T., de Alvarenga, S.C., A survey of intrusion detection in Internet of Things. *J. Netw. Comput. Appl.*, 84, 25–37, 2017.
4. Trnka, S., Cerny, T., Stickney, N., Survey of authentication and authorization for the Internet of Things. *Secur. Commun. Netw.*, 2018, 1–17, 2018.
5. Kuila, S., Dhanda, N., Joardar, S., Neogy, S., Kuila, J., A generic survey on medical big data analysis using Internet of Things, in: *First International Conference on Artificial Intelligence and Cognitive Computing*, pp. 265–276, Springer, 2019.
6. Challoner, A. and Popescu, G.H., Intelligent sensing technology, smart healthcare services, and Internet of Medical Things-based diagnosis. *Am. J. Med. Res.*, 6, 1, 13–18, 2019.
7. Kang, S., Baek, H., Jung, E., Hwang, H., Yoo, S., Survey on the demand for adoption of Internet of Things (IoT)-based services in hospitals: Investigation of nurses' perception in a tertiary university hospital. *Appl. Nurs. Res.*, 47, 18–23, 2019.
8. Adhikary, T., Jana, A.D., Chakrabarty, A., Kumar Jana, S., The Internet of Things (IoT) augmentation in healthcare: An application analytics, in: *International Conference on Intelligent Computing and Communication Technologies*, pp. 576–583, Springer, India, 2019.
9. Kocabas, O., Soyata, T., Aktas, M.K., Emerging security mechanisms for medical cyber physical systems. *IEEE/ACM Trans. Comput. Biol. Bioinform.*, 13, 3, 401–416, 2016.
10. Venkatasubramanian, K.K., Banerjee, A., Gupta, S.K.S., PSKA: Usable and secure key agreement scheme for body area networks. *IEEE Trans. Inf. Technol. Biomed.*, 14, 1, 60–68, 2010.
11. Ye, F., Qian, Y., Hu, R.Q., Energy efficient self-sustaining wireless neighborhood area network design for smart grid. *IEEE Trans. Smart Grid*, 6, 1, 220–229, 2015.
12. Koydemir, H.C. and Ozcan, A., Wearable and implantable sensors for biomedical applications. *Annu. Rev. Anal. Chem.*, 11, 127–146, 2018.
13. Hiremath, S., Yang, G., Mankodiya, K., Wearable Internet of Things: Concept, architectural components and promises for person-centered healthcare, in: *2014 4th International Conference on Wireless Mobile Communication and Healthcare-Transforming Healthcare Through Innovations in Mobile and Wireless Technologies*, pp. 304–307, MOBIHEALTH, IEEE, Greece, 2014.
14. S.G.J. Yuen, H. Park, A. Ghoreyshi, A. Wu, User identification via motion and heartbeat waveform data, Google Patents, US Patent no. US20170035327A1, 2017.

15. Pinto, M.B. and Yagnik, A., Fit for life: A content analysis of fitness tracker brands use of Facebook in social media marketing. *J. Brand Manag.*, 24, 1, 49–67, 2017.
16. Asmar, R., Validation of the automatic blood pressure measurements device, the OMRON EVOLV (HEM-7600 TE)® in pregnancy according to the modified European Society of Hypertension International Protocol (ESH-IP). *Med. Res. Inst.*, 2017, 1–10, 2017.
17. Wang, R., Blackburn, G., Desai, M., Phelan, D., Gillinov, L., Houghtaling, P., Gillinov, M., Accuracy of wrist-worn heart rate monitors. *Jam. Cardiol.*, 2, 1, 104–106, 2017.
18. Khezr, S., Moniruzzaman, Md., Yassine, A., Benlamri, R., Blockchain Technology in Healthcare: A Comprehensive Review and Directions for Future Research. *Appl. Sci.*, 9, 1736, 2019.
19. Antonakakis, M. *et al.*, Understanding the Mirai Botnet, in: *26th USENIX Conference on Security Symposium, SEC'17*, vol. 2017, pp. 1093–1110, 26th USENIX Security Symposium, Berkeley, CA, 2017.
20. Mirzamohammadi, S. and Sani, A.A., The case for a virtualization-based trusted execution environment in mobile devices, in: *Asia-Pacific Workshop on Systems*, pp. 1–8, NY, 2018.
21. Herder, C., Yu, M., Koushanfar, F., Devadas, S., Physical unclonable functions and applications: a tutorial. *Proc. IEEE*, 102, 8, 1126–1141, 2014.
22. Zhang, Z., Cho, M.C.Y., Wang, C., Hsu, C., Chen, C., Shieh, S., IoT security: Ongoing challenges and research opportunities, in: *7th Int. Conf. on Service-Oriented Computing and Applications*, pp. 230–234, 2014.
23. Atzori, L., Iera, A., Morabito, G., The internet of things: A survey. *Comput. Netw.*, 54, 15, 2787–2805, 2010.
24. Cheng, K., Bai, Y., Wang, R., Ma, Y., Optimizing soft real-time scheduling performance for virtual machines with SRT-Xen, in: *15th International Symposium on Cluster, Cloud and Grid Computing (CCGrid)*, pp. 69–178, 2015.
25. Balasamy, K. and Ramakrishnan, S., An intelligent reversible watermarking system for authenticating medical images using wavelet and PSO. *Clust. Comput.*, 22, 2, 4431–4442, 2019.
26. Bellard, F., QEMU, a fast and portable dynamic translator, in: *Annual Conference on USENIX Annual Technical Conference*, p. 41, Berkeley, CA, 2005.
27. Barham, P., Dragovic, B., Fraser, K., Hand, S., Harris, T., Ho, A., Neugebauer, R., Pratt, I., Warfield, A., Xen and the art of virtualization. *SIGOPS Oper. Syst. Rev.*, 37, 5, 164–177, 2003.
28. Habib, I., Virtualization with, K.V.M., *Linux J.*, 166, 2008.
29. Dall, C. and Nieh, J., KVM/ARM: The Design and Implementation of the Linux ARM Hypervisor, in: *19th International Conference on Architectural Support for Programming Languages and Operating Systems*, pp. 333–348, New York, NY, 2014.

30. Yoo, S. and Yoo, C., Real-time scheduling for Xen-arm virtual machines. *IEEE Trans. Mob. Comput.*, 13, 8, 1857–1867, 2014, https://doi.org/10.1109/TMC.2013.109.
31. Avanzini, A., Valente, P., Faggioli, D., Gai, P., Integrating Linux and the real-time ERIKA OS through the Xen hypervisor, in: *Int. Symp. Industrial Embedded Systems (SIES)*, pp. 1–7, 2015.
32. Balasamy, K. and Suganyadevi, S., A fuzzy based ROI selection for encryption and watermarking in medical image using DWT and SVD. *Multimed. Tools Appl.*, *80*, 7167–7186, 2021, https://doi.org/10.1007/s11042-020-09981-5.
33. Trujillo, S., Crespo, A., Alonso, A., MultiPARTES: multicore virtualization for mixed criticality systems, in: *Euromicro Conference on Digital System Design (DSD)*, pp. 260–265, 2013.
34. Zhou, R., Ai, Z., Yang, J., Chen, Y., Li, J., Zhou, Q., Li, K.-C., A hypervisor for MIPS-based architecture processors—A case study in Loongson processors, in: *International Conference on High Performance Computing and Communications*, pp. 865–872, 2013.
35. Patel, A., Daftedar, M., Shalan, M., Watheq El-Kharashi, M., Embedded Hypervisor Xvisor: A Comparative Analysis, in: *23rd Euromicro International Conference on Parallel, Distributed and Network-Based Processing (PDP)*, pp. 682–691, 2015.
36. Xia, T., Prevotet, J.-C., Nouvel, F., Mini-NOVA: a lightweight ARM-based virtualization microkernel supporting dynamic partial reconfiguration, in: *International Parallel and Distributed Processing Symposium Workshop (IPDPSW)*, pp. 71–80, 2015.
37. Heiser, G. and Leslie, B., The OKL4 microvisor: convergence point of microkernels and hypervisors, in: *Asia-Pacific Workshop on Workshop on Systems*, NY, pp. 19–24, 2010.
38. Balasamy, K. and Shamia, D., Feature Extraction-based Medical Image Watermarking Using Fuzzy-based Median Filter. *IETE J. Res.*, 2021, https://www.tandfonline.com/doi/full/10.1080/03772063.2021.1893231.
39. Kshetri, N., Can Blockchain strengthen the internet of things? *IT Prof.*, 19, 4, 68–72, 2017.
40. Johnson, D., Menezes, A., Vanstone, S., The elliptic curve digital signature algorithm (ECDSA). *Int. J. Inf. Secur.*, 1, 1, 36–63, 2001.
41. Gilbert, H. and Handschuh, H., Security analysis of SHA-256 and sisters, in: *Selected Areas in Cryptography*, M. Matsui and R.J. Zuccherato (Eds.), pp. 175–193, Springer, Berlin, Heidelberg, 2004.
42. Wang, H., Zheng, Z., Xie, S., Dai, H.-N., Chen, X., Blockchain challenges and opportunities: A survey. *Int. J. Web Grid Serv.*, 14, 352, 2018.
43. Banafa, A., IoT and Blockchain convergence: Benefits and challenges. *IEEE Internet Things*, 2017, 1–9, 2017.
44. Krishnasamy, B., Balakrishnan, M., Christopher, A., A Genetic Algorithm Based Medical Image Watermarking for Improving Robustness and Fidelity in Wavelet Domain, in: *Intelligent Data Engineering and Analytics*. Advances

in Intelligent Systems and Computing, vol. 1177, Springer, Singapore, 2021, https://doi.org/10.1007/978-981-15-5679-1_27.
45. Dorri, A., Kanhere, S.S., Jurdak, R., Gauravaram, P., Blockchain for IoT security and privacy: The case study of a smart home, in: *International Conference on Pervasive Computing and Communications Workshops (PerCom Workshops)*, pp. 618–623, 2017.
46. Dorri, A., Kanhere, S.S., Jurdak, R., Towards an Optimized BlockChain for IoT, in: *IEEE 2nd Int. Conf. on Internet-of-Things Design and Implementation*, pp. 173–178, New York, NY, 2017.
47. Brezillon, P. and Mostefaoui, G.K., Context-based security policies: A new modelling approach, in: *Conference on Pervasive Computing and Communications Workshops*, pp. 154–158, 2004.
48. Perera, C., Zaslavsky, A., Christen, P., Georgakopoulos, D., Context aware computing for the internet of things: A survey. *IEEE Commun. Surv. Tutorials*, 16, 1, 414–454, 2014.
49. Abowd, G.D., Dey, A.K., Brown, P.J., Davies, N., Smith, M., Steggles, P., Towards a better understanding of context and context-awareness, in: *1st International Symposium on Handheld and Ubiquitous Computing*, vol. 1999, pp. 304–307, Springer-Verlag, London, 1999.
50. Ramos, J.L.H., Bernabe, J.B., Skarmeta, A.F., Managing context information for adaptive security in IoT environments, in: *29th International Conference on Advanced Information Networking and Applications Workshops*, 2015, 676–681, 2015.
51. Habib, K. and Leister, W., Context-aware authentication for the internet of things, in: *Eleventh International Conference on Autonomic and Autonomous Systems*, 2015, 6, 2015.
52. Trnka, M. and Cerny, T., On security level usage in context-aware role-based access control, in: *31st Symposium on Applied Computing*, vol. 2016, pp. 1192–1195, New York, NY, 2016.
53. Hayashi, E., Das, S., Amini, S., Hong, J., Oakley, I., CASA: Context-aware scalable authentication, in: *Ninth Symposium on Usable Privacy and Security*, vol. 10, pp. 3:1–3:10, New York, NY, 2013.
54. Hosseinzadeh, S., Virtanen, S., Díaz-Rodríguez, N., Lilius, J., A semantic security framework and context-aware role-based access control ontology for smart spaces, in: *International Workshop on Semantic Big Data*, vol. 8, pp. 1–8, New York, NY, 2016.
55. Langheinrich, M., *A Privacy Awareness System for Ubiquitous Computing Environments*, pp. 237–245, Springer, Berlin, Heidelberg, Chapter 19, 2002.
56. Pingley, A., Yu, W., Zhang, N., Fu, X., Zhao, W., A context-aware scheme for privacy preserving location-based services. *Comput. Netw.*, 56, 11, 2551–2568, 2012.
57. Pinto, S. *et al.*, IIoTEED: An enhanced, trusted execution environment for industrial IoT edge devices. *IEEE Internet. Comput.*, 21, 40–47, 2017.

58. Jang, J.S. *et al.*, SeCReT: secure channel between rich execution environment and trusted execution environment, in: *Network and Distributed System Security Symposium (NDSS)*, 2015, 1–15, 2015.
59. Rifi, N., Rachkidi, E., Agoulmine, N., Taher, N.C., Towards using blockchain technology for IoT data access protection, in: *17th International Conference on Ubiquitous Wireless Broadband (ICUWB)*, pp. 1–5, 2017.
60. Polyzos, G.C. and Fotiou, N., Blockchain-assisted information distribution for the Internet of Things, in: *International Conference on Information Reuse and Integration (IRI)*, pp. 75–78, 2017.
61. Ouechtati, H., Azzouna, N.B., Said, L.B., Towards a self-adaptive access control middleware for the Internet of Things. *Int. Conf. on Information Networking (ICOIN)*, pp. 545–550, 2018.

8

Blockchain in Smart Healthcare Management

Jayant Barak, Harshwardhan Chaudhary, Rakshit Mangal, Aarti Goel* and Deepak Kumar Sharma

Department of Information Technology, Netaji Subhas University of Technology, New Delhi, India

Abstract

Blockchain is one of the latest and emerging technology advances that facilitate verifiable activities between parties reliably and rapidly. It builds on the idea of a P2P (Peer-To-Peer) network and provides a decentralized, distributed, shared, and immutable database ledger where block data contain a list of all transactions linked by cryptographic hashes. Blockchain eliminates the need for a legitimate central supervisory authority. It uses the state of art consensus and cryptographic algorithms. When new transactions are added to Blockchain, these algorithms are used to verify data's authenticity in each node. Blockchain technology also makes it difficult to directly alter a part of the blockchain as the hashes of subsequent nodes of the altered node must be recomputed. This requires infeasible computational power, making the security fool-proof and increasing exponentially with the chain's size. The application of Blockchain in such a manner makes it easier to verify the legitimacy of transactions and at the same time provide transparency, reliability, and integrity. Blockchain is implemented in a wide range of domains such as Supply chain management (SCM), financial sector, Internet of Things (IoT), Energy Sector, and Healthcare. The rapid growth of digitization in healthcare has paced the generation of massive electronic records of patients. This calls for protecting such highly confidential data and ensuring data integrity during transactions between authorized parties. Blockchain in healthcare proposes a mechanism that is transparent and responsible. It can revolutionize healthcare by providing patients' privacy, allowing them to retain ownership

**Corresponding author*: aartigoelcse@gmail.com

Kavita Sharma, Yogita Gigras, Vishnu Sharma, D. Jude Hemanth and Ramesh Chandra Poonia (eds.) Internet of Healthcare Things: Machine Learning for Security and Privacy, (161–188) © 2022 Scrivener Publishing LLC

of their records, simultaneously giving them easy access to hospitals. In this chapter, applications of Blockchain in a smart healthcare system are explored. Blockchain in Healthcare Systems can be securely implemented in sharing, storing and creating electronic medical records and prescriptions, personalized medicine, remote monitoring, mobile health applications and the Internet of Medical Things (IOMT) vital to patient care and its improved quality. Fraud detection, Insurance claims, medical transactions, checking Drug Counterfeiting, and Tracking, along with other utilities for patients, doctors, and healthcare institutes, can also be achieved using blockchain in healthcare. Other useful applications include Public Health Management and supervising clinical trials. Blockchain can be used to ensure the credibility of information related to medical resources available online. Various Machine Learning (ML) and deep learning algorithms can also be applied to data to improve treatment effectiveness and efficiency.

Keywords: Blockchain, cryptography, hashing, consensus algorithm, healthcare, internet of medical things (IOMT), machine learning (ML), green blockchain

8.1 Introduction

Blockchain is one of the latest and emerging technology advances that facilitate verifiable activities between parties reliably and rapidly. It is suggested by Forecast that global revenue generated by blockchain technology is projected to have massive growth, with the market surpassing 39 billion USD by 2025. Blockchain builds on the idea of a P2P (Peer-To-Peer) network. Its architecture eliminates the need for a central supervising authority. Isn't it amusing to explore a future where there is no need to depend on a third party for their services? Many apps adopting decentralized P2P platforms (DApps or Decentralized Apps) where users and suppliers will get connected without getting controlled by any third party. Startup DApps are giving Uber and Amazon a run for their money. Blockchain is widely used in cryptocurrencies such as Bitcoin, Ethereum, etc. The most common hurdle faced in Healthcare is having access to all medical records in one place. They are generally spread among various healthcare providers and stakeholders. Integration of Blockchain with healthcare can address these issues along with other challenges that are being discussed in this chapter. Blockchain in healthcare proposes a mechanism that is transparent and responsible. It can revolutionize healthcare by ensuring patients' privacy, allowing them to retain ownership of their records, simultaneously giving easy access to hospitals.

The main aim is to achieve effective data management with efficient sharing and processing while safeguarding privacy and security when dealing with stakeholders.

In this chapter, Healthcare systems and their challenges, followed by Blockchain and its integration with Healthcare are discussed. Blockchain in Healthcare Systems can be securely implemented in sharing, storing, and creating electronic medical records and prescriptions, personalized medicine, remote monitoring, mobile health applications, and the Internet of Medical Things (IoMT) which is vital to patient care and improves its quality. Fraud detection, Insurance claims, Medical transactions, checking Drug Counterfeiting, and Tracking, along with other utilities for patients, doctors, and healthcare institutes, can also be achieved by using blockchain in healthcare. Other useful applications include Public Health Management and supervising clinical trials. Various Machine Learning (ML) and deep learning algorithms can also be applied to data to improve the effectiveness and efficiency of treatment. Also discussed are the challenges faced while implementing blockchain in healthcare.

8.2 Healthcare Industry

The Healthcare Industry is vast and has various sectors for providing healthcare facilities and products for patients. The interaction between different Healthcare stakeholders involve parties like the patient, the Doctor, the Hospital staff and Management, the Pharmaceutical company, the Insurance company, or the Government. Such parties and all the intermediaries need to have effective and error-free data conveyance for reliability and to avoiding fraudulency.

Healthcare mainly concentrates on maintenance and health improvement by prevention, management, diagnosis, treating diseases, illness, injuries, and other physical and mental ailments. Healthcare availability to the public varies across different countries and communities and is largely driven by the social and economic environment and government policies. However, healthcare organizations are established to meet the health needs of the general public [1].

8.2.1 Classification of Healthcare Services

The provision of healthcare services to the public is classified into the following three types as shown in Figure 8.1.

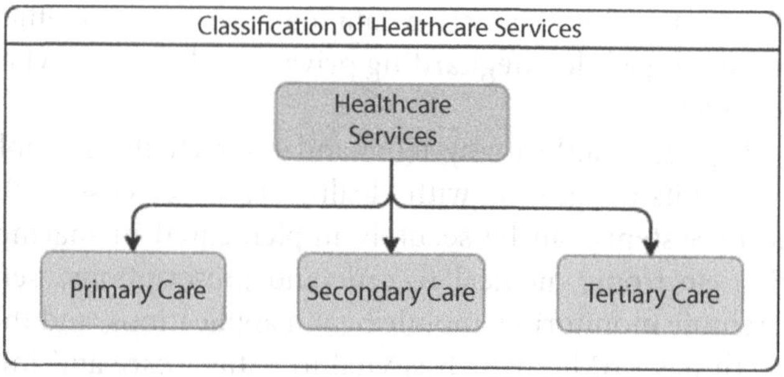

Figure 8.1 Classification of healthcare services.

1) **Primary Care**: This is the first healthcare-related service received by a patient. These services usually consist of a general practitioner/family physician who acts as the first contact for the patient in the treatment process. The Primary Care doctor should be able to accurately diagnose the patient based on his broad sense of knowledge.
2) **Secondary Care**: This primarily involves acute care, usually given in hospital emergency departments. Here only essential treatment is given and if necessary, the patient is referred to a tertiary health department. Therefore, the care given by a secondary physician is usually for a brief window of time.
3) **Tertiary Care**: This is a specialized healthcare-related service usually given to patients on referral from primary or secondary health departments. This sector requires more care and time to heal the illness, usually for complex medical and surgical processes.

8.2.2 Health Information Technology (HIT)

Before studying the Healthcare sector and its stakeholders, it is essential to know about HIT or Health IT and its components related to this chapter for better understanding later on. Health Information Technology (HIT) involves the processing and management of healthcare data that leads to efficient and effective communication and decision making [2].

Various components of HIT are:

- **Electronic Health Record (EHR)**—EHR stores extensive medical history of patients and organizes records from multiple providers together in one place.

- **Electronic Medical Record (EMR)**—EMR stores standard clinical and medical information of a patient, all gathered in the healthcare provider's office.
- **Personal Health Record (PHR)**—PHR is a privately maintained medical history of patients for personal use only.
- **Medical Practice Management software (MPM)**—It is a management software that is designed to facilitate everyday operations in a medical facility efficiently.
- **Health Information Exchange (HIE)**—HIE allows patients to securely share their confidential medical information electronically with healthcare providers without compromising their privacy.

8.2.3 Issues and Challenges Faced by Major Stakeholders in the Healthcare Industry

The Healthcare industry's major stakeholders (as depicted in Figure 8.2) are the Patient, the Healthcare service provider, the Pharmaceutical Company, an Insurance company, and the Government. These are discussed in detail below.

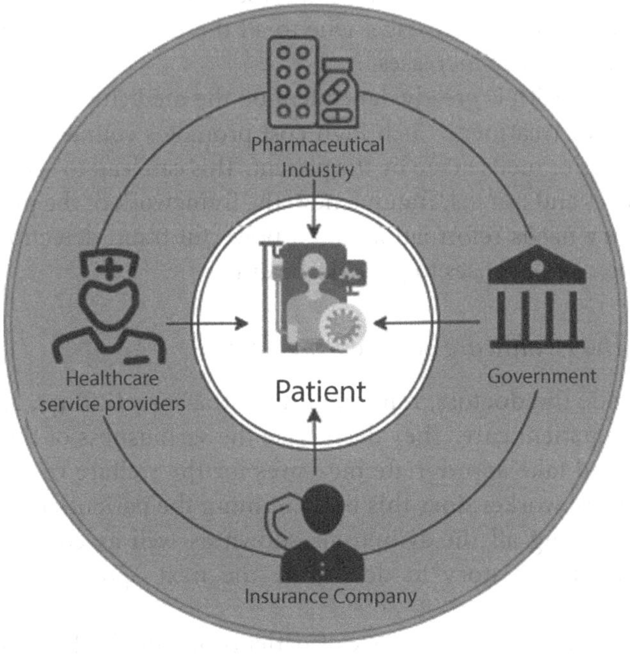

Figure 8.2 Stakeholders in healthcare.

8.2.3.1 *The Patient*

The Patient is the end receiver of services provided in the healthcare industry and becomes a significant stakeholder. The only way of assessing if the care and facility provided were up to the mark depends upon the patient's satisfaction and the recovery made in the treatment. Patient satisfaction depends mainly on the quality of service received and expenses incurred during the process.

This all should be measured while the information remains private or is shared only to the authorized parties, and there is transparency in the transactions being made. The presence of centralized management of healthcare databases makes it vulnerable to cybercrimes like data theft, manipulation, and misuse. This calls for a secure data flow and permission granting to authorize parties in the process, ensuring privacy and correct use of patient information.

8.2.3.2 *The Pharmaceutical Industry*

The role of pharmaceutical companies is to manufacture medicines or drugs which are required by a patient for his/her treatment, thus playing an important role in the healthcare sector. Before manufacturing medical drugs, the company has to invest in its research and development before the mass-scale production stage. Due to all these intermediary steps, the cost of a medical drug increases.

A patient's health is greatly dependent on the medicines or drugs taken by him/her for treatment. Such high cost promotes counterfeiting of the medical drugs or meds taken by the patient. This can lead to serious health consequences and not just fraudulency. The framework of the pharmaceutical industry needs reformation and a powerful fraud detection system, ensuring no compromises to patients' health.

8.2.3.3 *The Healthcare Service Providers*

These include the doctors, nurses, or any other practitioner, and play a vital role in patient care. They determine the seriousness of the patient's condition and take appropriate measures for the welfare of the patient. The healthcare worker does this by examining the patient's current condition, measuring all the symptoms shown as well as considering the patient's medical history to determine the next appropriate step for treatment.

With the increase in the influx of patients, it would take a lot of time to compile the results of a series of tests conducted on patients, just to diagnose patients' actual problems and suggest a befitting treatment. The compilation step itself takes a lot of time because data from various tests are produced in a fragmented and decentralized way. And the transfer of this data from one authorized party to another might even result in incorrect data updating if no proper system is used. This calls for efficient interoperability and a secure network among all the decentralized systems so that no tampering is done with the patients' reports.

8.2.3.4 *The Government*

The government plays an important role in the whole healthcare industry due to its influence. It includes financial and material resources for Medical institutions, funding the development of human resources and research opportunities in public health services. Other roles of government include implementation of health-related policies, enforcing hygiene and safety standards on various institutions and industries, promoting health via different initiatives, and creating awareness. Its functioning and policies are prone to faults and loopholes owing to the scale of its policies introduced and the number of people involved [3].

8.2.3.5 *Insurance Company*

Health insurance covers the whole or a part of the cost for the services and prescription a person took for his/her treatment. It is a binding contract between health insurance providers and their policyholders. This contract usually involves monthly/annual payments of premiums. The Health Insurance provider cites in writing the type and amount of healthcare costs that will be covered under the policy. All these are provided by an Insurance company [4].

Such insurance systems have problems such as poor access to the patient data thus assignment of wrong policies, manipulation of details of the patient's record leading to the incurrence of inaccurate insurance claims. These call for a consistent database of records, which is immutable, assuring the Insurance Company in the insurance they provide is accurate and adequate for the treatment. This would develop trust amongst the Insurance Company, the Healthcare Service provider and the Patient.

Blockchain integrated into the Healthcare system seems to provide solutions to the above-mentioned issues faced by the stakeholders. Now focus on blockchain technology and its working before and explore the healthcare aspect of it.

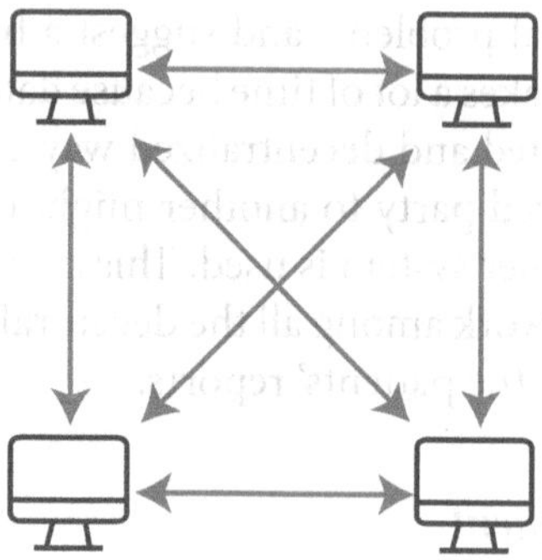

Figure 8.3 P2P network.

8.3 Blockchain Technology

The Blockchain is in its most trivial is a linked list that contains time-stamped series of immutable data connected by a hash of the previous block. Blockchain is managed by a cluster of computers called nodes, which maintain a copy of the chain. Each block in a blockchain is secure, immutable and connected to the previous block via its hash [5]. It builds on the idea of a P2P network where P2P stands for "Peer to Peer."

A P2P network consists of computers called "peers" that are connected via the internet as shown in Figure 8.3. Files can be directly shared between peers without a central server [6]. Hence, each peer acts as a file server as well as a client [6]. In this way, the P2P network provides a decentralized, distributed, shared, and immutable database ledger where block data contain a list of all transactions by eliminating the need for a legitimate central supervisory authority.

8.3.1 Important Terms

1. **Block:** Blockchain comprises different blocks. These Blocks contain data, nonce, and the hash of the current block, which stores the output of the hash function, which takes in data. Nonce, the hash of the previous block, timestamp as input. It also stores the hash of the previous block as represented in Figure 8.4.
2. **Miners:** Miners create new blocks and add them to Blockchain. Since every block refers to previous blocks, Miners solve incredibly complex math problems to find a nonce that generates an accepted hash using special software.

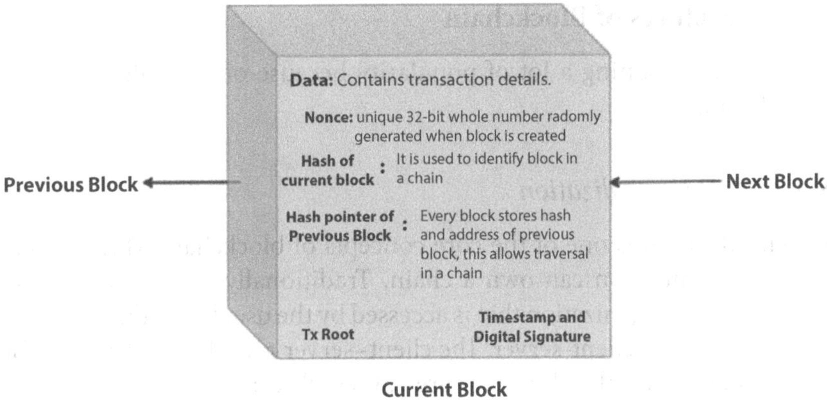

Figure 8.4 Block in blockchain.

3. **Nonce:** Nonce is usually a 32-bit whole number that is randomly generated while mining a block. The hash function takes data and nonce as input. The nonce is treated as a variable parameter, which is randomly changed during the mining process to reach a hash that adheres to the rules of an acceptable.
4. **Nodes:** Nodes contribute to the distributive ledger property of Blockchain. They are electronic devices that store and maintain copies of blockchain and keep the network running. This network should consensually approve newly mined blocks to add new blocks to the chain [7].
5. **Timestamp:** It is the total elapsed time in seconds since UTC 00:00:00 on January 1, 1970 [8].
6. **Consensus:** It involves verification and validation of information in a distributed ledger system by eliminating the need to rely on a central authority [13]. Popular consensus algorithms used by most Blockchain architectures nowadays are Proof of Work (PoW), Delegated Proof of Stake (DPoS), or Proof of Stake (PoS). These algorithms ensure that all the nodes have an untampered copy of data and new blocks added to the chain are genuine [9].
7. **Tx_Root:** It is also referred to as the Merkle root [21]. It contains the hash value of all validated transactions of the block. All the hashes of the blocks are combined pairwise with each other and are given as input to another hash function. This work is repeated until there is only a single hash, which stands for the Merkle root [10].

8.3.2 Features of Blockchain

Blockchain is gaining a lot of popularity because of the following mentioned features.

8.3.2.1 Decentralization

Decentralization is one of the core concepts of blockchain which implies that no organization can own a chain. Traditionally, data is stored on a database of an organization that is accessed by the user by sending a request to a server called client-server. The client–server gives back relevant information [23]. Centralized systems are vulnerable as:

1. They are targets for potential data hacks since all the data of the organization is stored at one spot.
2. During maintenance or in case of system failure, the data becomes inaccessible.
3. In the worst-case scenario, the centralized system could get corrupted or infected with malware. In that case, all of the data residing in the centralized system will be compromised [5].

Blockchain is based on a P2P network where everyone has copies of the chain. This eliminates the need for a third party to enable a transaction and provide users with a copy of their information. It also avoids a single point of failure situation [8].

8.3.2.2 Immutability

Data/Block once added to the blockchain, cannot be tampered with. This is attained via the use of cryptographic hash functions, which take the data from the block as an input. Hashing consists of taking an input string of variable length to give out an output of fixed length. Input data can also be tracked using the hash of the block.

The "Avalanche Effect" property [5] of cryptographic hash functions plays a huge role in ensuring immutability. This property of cryptographic hash functions reflects huge changes in output hash even for small changes in input. If someone tries to change the block's data, even the slightest changes will change the hash drastically. The nonce and hash are 32 and 256 bits respectively. Thus, there are roughly four billion possible nonce-hash combinations that must be mined before the right one is found. When our miner finds the right combination, the mined nonce is called "golden nonce". The miner is awarded a reward and the block is added to the chain.

Blockchain technology makes it difficult to directly alter a part of the blockchain as the hashes of subsequent nodes of the altered node have to be recomputed and the blocks have to be mined again. Relying on "safety in math" as finding the right nonce requires huge amounts of time and computing power [7]. This avalanche effect makes the security fool-proof and increases it exponentially with the size of the chain. The application of Blockchain in such a manner makes it easier to verify the legitimacy of transactions.

8.3.2.3 Transparency

User's identity and confidential data are hidden using complex cryptographic techniques and represented by their public-key/public address. In this way, Blockchain provides data security yet at the same time provide transparency about transactions [22].

8.3.2.4 Smart Contracts

Smart contracts are executable codes that enforce appropriate rules and policies on different authorized parties to facilitate their interaction with each other and the data. Smart contracts enable automation when specified conditions are met. Thus, the use of a Smart contract assures a low cost of implementation compared to current management systems due to the elimination of previously required trusted third parties for enforcement [11].

8.3.3 Workings of a Blockchain System

Blockchains are roughly categorized into three different types:

- Public blockchain;
- Private blockchain; and
- Consortium blockchain.

In a public blockchain, every participant can become a node and participate in the consensus process and all the transactions/records are visible publicly. On the other hand, in a Consortium blockchain, only a group of pre-selected nodes can participate in the consensus process. In a private blockchain, only nodes that are associated with specific organizations can participate in the consensus process [8].

In a blockchain system, each user owns a pair of public key and a private key. The private key of the user is kept confidential and is used to perform digital signatures. Blockchain's work can be easily understood with help of Figure 8.5. First, a user digitally signs a transaction that he participated in and then makes a transaction request which is broadcasted to the P2P network.

The digital signature consists of two important phases: the signing phase and the verification phase. Suppose the sender wants to send a message to a receiver. While signing, the sender will encrypt data using their private key and then sends the receiver the message. While verifying, the receiver validates the data using the public key of the sender. This protocol allows the receiver to easily check if data is tampered with or not. One of the most widely used algorithms to perform digital signatures is the Elliptic curve digital signature algorithm [8].

After validation and verification of the transaction, a new data block is created. This block is then mined and added to the chain. The Blockchain is in its most trivial is a linked list that contains time-stamped series of immutable data connected by a hash of the previous block. The nonce is used to generate a hash which is acceptable as discussed before. The data in the block is considered signed and forever tied to nonce and hash [7]. Any changes in data,

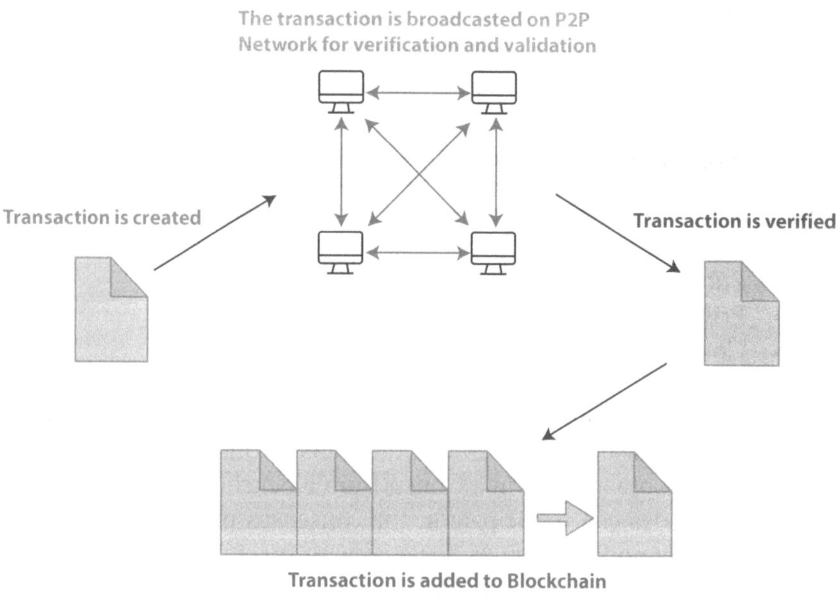

Figure 8.5 Working of a blockchain system.

even small changes, will trigger the before avalanche effect discussed earlier and all the subsequent nodes have to be mined again.

8.3.4 Applications of Blockchain

8.3.4.1 Financial Services

Funds transfer has always been observed to be slow and costly using the traditional modes of transactions in centralized systems. But the Blockchain provided the feature of Distributed Ledger Systems which made lending and borrowing of money speedy, reliable and transparent across all the banks and borders [1].

8.3.4.2 Healthcare

Healthcare systems contain a huge amount of data and in case any critical information gets lost or altered then it may cost the life of the patient. Thus, Blockchain provides a feature of Distributed Ledger Systems (DLT) and immutability in which the data and information are secure and unalterable.

8.3.4.3 Supply Chain

With the tremendous increase in the manufacturing processes around the globe, the need for transparency between the supplier and the supply chains is a must. Distributed Ledger Systems (DLT) in Blockchain offers end-to-end decentralized processes that make its system transparent. Blockchain is perfectly suitable for supply chain management, with real-time tracking of goods and especially appealing to companies having multiple supply chains.

8.3.4.4 Identity Management

Blockchain helps in tracking and managing digital identities securely and efficiently which results in the reduction of data theft and fraud. Verification and authentication are nowadays required in every industry whether it be insurance, healthcare, banking, national security, etc.

8.3.4.5 Voting

The voting systems at present are inefficient and prone to manipulation. Blockchain technology can improve this system to identify the rationality of individual citizens. Distributed Ledger Systems (DLT) used to store the voting data means that the result is not managed by a centralized authority thus eliminating the menace of voting result manipulations [1].

8.3.5 Challenges and Drawbacks of Blockchain

Although the technology seems very promising yet it faces some challenges. These challenges may be resolved in the future with innovative solutions and emerging technologies. Now we discuss some of the challenges based on our current technological infrastructure (as shown in Figure 8.6).

Scalability:
Scalability is a major drawback of Blockchain Technology. When one uses the Bitcoin Network, the transactions completed depend on the network congestion i.e. the more the number of individuals or nodes the more is the possibility of the system slowing down. For instance, Visa and Bitcoin can process 65K and seven transactions per second respectively.

High Energy Consumption:
High energy consumption is a very concerning drawback especially when the world is moving towards a green and clean future. The consumption

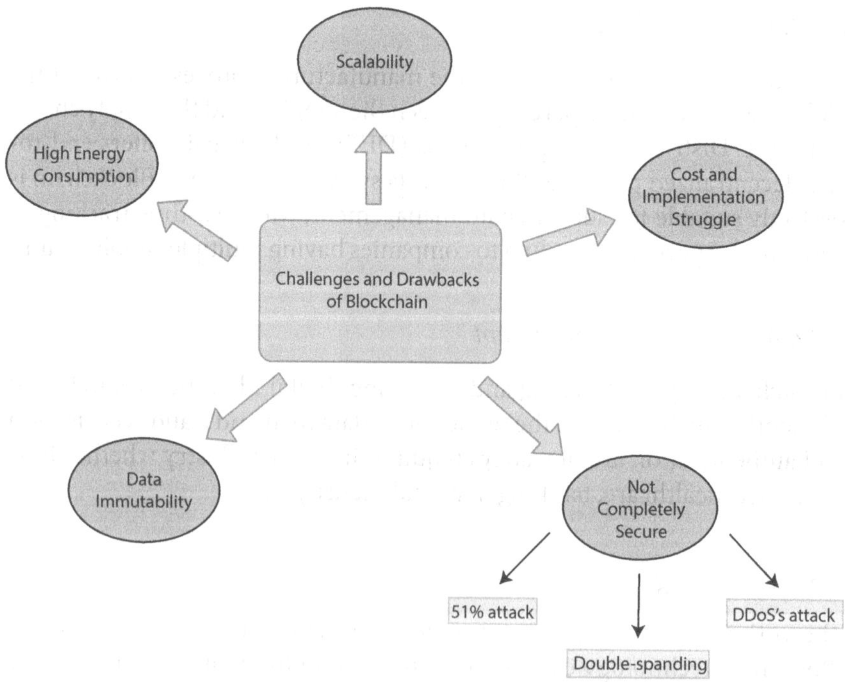

Figure 8.6 Drawbacks of blockchain.

of power in the Blockchain is needed for keeping a real-time track of the transactions [13]. Whenever a new node is created, the miners find a valid nonce once the transactions are validated by the network, thus spending a lot of energy. It has been estimated by the Researchers at the University of Cambridge that Bitcoin consumes more energy than Switzerland [12, 14]. Now need to find ways to reduce this high energy consumption or else this consumption will only increase with the growth of blockchain platforms.

Data Immutability:
The data once written in the nodes cannot be altered or removed. Immutability should only be ensured if the network nodes are distributed uniformly among different organizations. A blockchain network (peer-to-peer) could be manipulated if a person/organization owns or possesses 50% or more nodes participating in the network, making the chain vulnerable.

Not Completely Secure:
Although Blockchain technology is safer than alternative platforms yet it is not 100% foolproof. There are ways in which Blockchain security could be compromised. Some of them are:

1. **51% attack:** In a 51% attack, if an organization can manage 51% or more of the network nodes, then it will take the management of the network. By doing so, they can alter the information within the transactions and additionally do double-spending.
2. **Double-spending:** Double spending can only be possible on the networks with an exposure to the 51% attack. Different consensus algorithms like Proof of Work (POW), Proof of Stake (POS), etc. can be used to prevent it.
3. **DDoS's attack:** During a DDoS attack, nodes of Blockchain are bombarded with similar kinds of requests which leads to congestion of the network, thereby crashing the entire network.

Cost and Implementation Struggle:
The cost for implementation, execution and maintenance of blockchain technology is quite high. Some prices include the hiring of developers, licensing cost if one opts for a paid blockchain solution, expenses of the management of a team that is an expert in various aspects of blockchain technology, and so on. The cost of the maintenance associated with the solution also needs to be taken care of.

Lastly, it can be concluded that blockchain is not quite ready for practical applications and there is the scope of improvement to make it more reliable for daily needs.

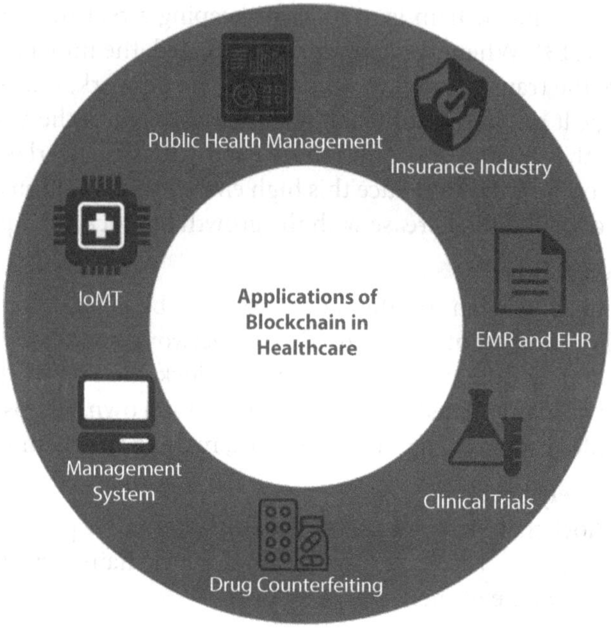

Figure 8.7 Applications of blockchain in healthcare.

8.4 Applications of Blockchain in Healthcare

Blockchain can be applied across various domains in Healthcare. Now explore the applications of Blockchain (refer to Figure 8.7) that shall give a boost to healthcare infrastructure catering to present needs and making it more efficient.

8.4.1 Electronic Medical Records (EMR) and Electronic Health Records (EHR)

As discussed earlier EHR contains comprehensive information of patients going beyond standard clinical data while EMR keeps track of patient's medical and treatment history. Earlier, the patient's data was stored in printed documents. People were responsible for keeping track of their medical history by storing their prescriptions. The usage of digital records made it easier for patients and doctors to access medical history. Healthcare professionals generally have to work with limited data. This hampers subsequent diagnosis and appropriate treatment steps. Researchers also struggle to find desired data for undertaking various studies [15].

A centralized database to store patient data to improve the efficiency of patient care and increase the overall efficiency of healthcare systems seemed like the solution. This database along with storing patient's medical history also shared the total health of patients. This system was promising and increased patient care remarkably.

There are concerns about such a centralized database system that is prone to hacking. Moreover, healthcare systems are decentralized and giving responsibility to an organization raises concerns about data breaches and data leaks. Blockchain addresses all these concerns. It provides a distributed database ledger that stores encrypted immutable information. Every patient has a private key and a public key. The most common way to share confidential data between authorized parties can be achieved by using the RSA public-key cryptosystem.

8.4.2 Management System

Efficient communication, easy access to patient medical history, personalized treatments and focusing on holistic views of patients are vital and would ensure better management and drastically improve healthcare outcomes and patient care. Managing healthcare data is critical and very challenging as it involves dealing with confidential sensitive data along with maintaining the trust and faith of patients that their data is secure.

The integration of blockchain with healthcare widens its scope by closely knitting together different sectors and stakeholders of the healthcare Industry which improves efficiency and reduces cost by introducing automation via Smart contracts.

As discussed earlier, smart contracts enforce appropriate rules and policies on different authorized parties when they interact among themselves.

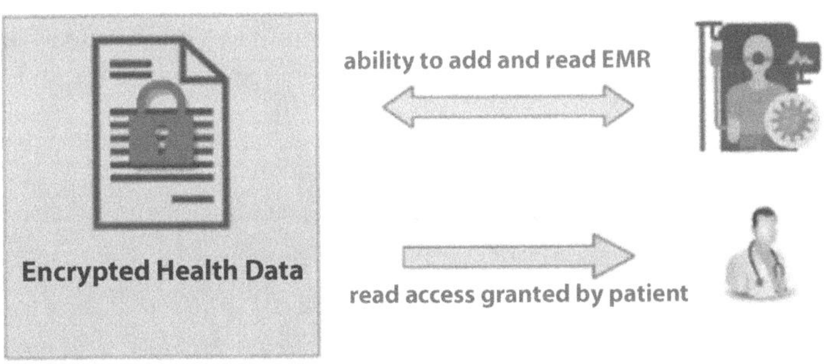

Figure 8.8 EMR (electronic medical records).

Laying down smart contracts between different stakeholders makes the system more efficient.

The patient doesn't have to file for an insurance claim; the bill can be generated automatically based on the insurance plan. This reduces cost, improves efficiency, and prevents insurance frauds and insurance claims evasion.

Researchers can have easy and accurate access to data for their studies which improves the quality of research and decreases research time at the same time ensuring privacy and correct use of patient information.

Epidemiologists can have access to data using which they can track the incidence and prevalence of different diseases in different areas and track the epidemic. The government can then communicate with these epidemiologists to put forward different programs and policies to control and potentially eradicate diseases. Blockchain with Healthcare could play a vital role in public health management.

8.4.3 Remote Monitoring/IoMT

Blockchain plays a vital role in sharing, storing and retrieval remote biomedical data generated by mobile devices, body sensors and IoMT (Internet of Medical Things) devices. Various applications of IoMT devices are discussed below:

1. Various data mining techniques, AI and deep learning algorithms can be used to keep track of health status, diagnose health conditions and prevent certain health complications thereby decreasing the duration of hospitalization which in turn decreases the workload of health professionals and saves money for patients.

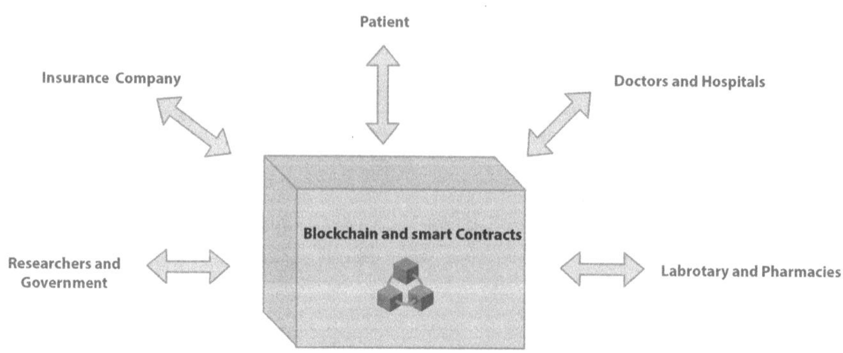

Figure 8.9 Healthcare management system using blockchain.

2. Chronic illnesses can be managed using IoMT devices. These devices have sensors that keep track of vitals and bodily systems. Routine tests can be scheduled and reported to the respective healthcare provider using portable diagnostic IoMT.
3. Along with having sensors that keep track of vitals and bodily systems, portable diagnostic IoMT devices can schedule routine tests and report the results to healthcare providers.
4. IoMT devices can help in ensuring that patients are taking their medicines regularly and as prescribed by their healthcare provider. It can do so by setting reminders for its regimen [16]. It can also be very helpful in reminding patients if they are running low on medicines.
5. Doctors can keep track of their prognosis of the disease and intervene if the patient doesn't recover within the expected recovery time.
6. Extensive contact tracing can be achieved with the help of IoMT devices which can help control epidemics helpful in the case of highly contagious diseases.

8.4.4 Insurance Industry

Blockchain Technology is bringing transformation in the insurance sector by introducing optimization in the business processes and sharing the information with better efficiency, security, and transparency. It is bringing the policy shift into the insurance system from manual to automated manner using the smart contracts on the peer-to-peer networks and thereby eliminating the normal processing system. There are numerous benefits that insurance companies and individuals seeking insurance can take using Blockchain Technology. With the help of decentralization, the insurance sector will get streamlined in underwriting, payments, claims,

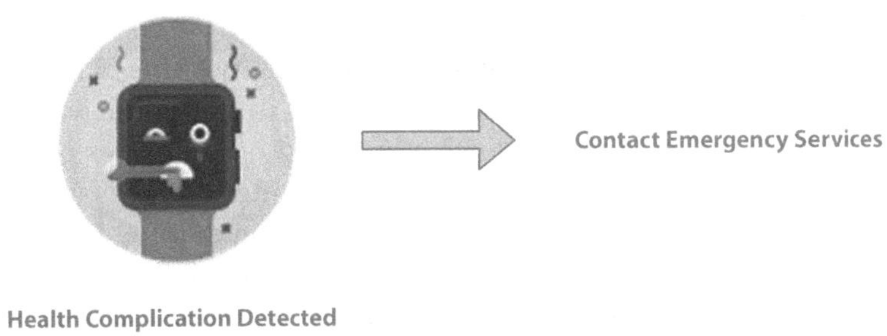

Figure 8.10 Remote monitoring using IoMT devices.

and reinsurance processes. This technology will provide higher security because the data is not stored at any centralized place and doesn't have any single entity control, thereby providing a better level of protection and cost-saving [1].

Without forcing insurers to conciliate patient data across various databases, Blockchain technology provides patients access and control over their medical data on a case by case basis. A cryptographic signature could be stored on a blockchain system for every record on a distributed ledger [18]. The signature creates an index of the data of the insurance documents in a cryptographical manner and timestamps it. This avoids storing any delicate information on the blockchain system.

As soon as some information is changed in an insurance document, it is registered on the shared ledger, which allows insurers to inspect the data across various organizations. Concurrently, the blockchain allows the granular permissions settings to adhere to the regulations while allowing data anonymization and sharing for research [18].

Today, the absolute complexness of the insurance industry creates gaps that are misused to commit fraud. It is during the slow, paperwork-driven claiming procedure having various moving parts, that the claims are transferred from Insurance-Policy holders to the insurers and reinsurers. Not only does it pave the way for brokers to sell insurance policies and make money off the premiums, but also allows criminals to make multiple claims from different insurance companies for a single loss.

With the implementation of Blockchain technology in the Insurance Industry, coordination between the insurers and the insurers can be made much more efficient, thus avoiding fraud [17]. Furthermore, registering the transactions on the distributed ledger with granular access controls would help maintain data security. Any abnormal and suspicious behaviour could be identified and would be taken care of appropriately by the insurance company as the claims would be recorded on the blockchain [18].

8.4.5 Drug Counterfeiting

A patient's health is largely dependent on the quality of medicinal drugs he/she takes. The research and production steps in the manufacturing of these drugs use a lot of capital. Such a high cost of medicines promotes counterfeiting. FDA defines drug/medicine counterfeiting as fake medicine that may be contaminated, contain the wrong dosage, and contain wrong or even no active ingredients in the medicine. They may contain toxic ingredients, which when consumed, can cause serious health issues.

Counterfeited drugs are introduced into the supply chain during its transportation or production stage where counterfeit drug manufacturers even use the brand logo of legitimate manufacturers to gain the trust of the consumer. World drug supply counterfeits are estimated to be from 10 to 15% of the total supply in first-world countries to up to 30% in developing countries. Due to the intake of the counterfeit drug, every year approximately 700,000 people die from tuberculosis and malaria. The WHO estimated that 1/5th of deaths that occur due to malaria is because of the counterfeit antimalarial drug [19].

Thus, the challenge faced by manufacturers is that their drug reaches the intended consumer in the original form and not from the counterfeiter. The Pharmaceutical industry needs a transparent Supply Chain Management (SCM) system so that consumers can be assured that the drug they're receiving for treatment is genuine. The SCM system when implemented with Blockchain ensures this. Figure 8.11 depicts the integration of Blockchain in the SCM system.

With Blockchain implementation, the manufacturer would be able to track and trace its product from manufacturing until it reaches the intended distributor. Data for that specific drug, which would be stored in a blockchain, will be immutable and it would be authenticated at each step. The consumer would be assured that the product received by him is legitimate because of Blockchain's tamper-proof nature.

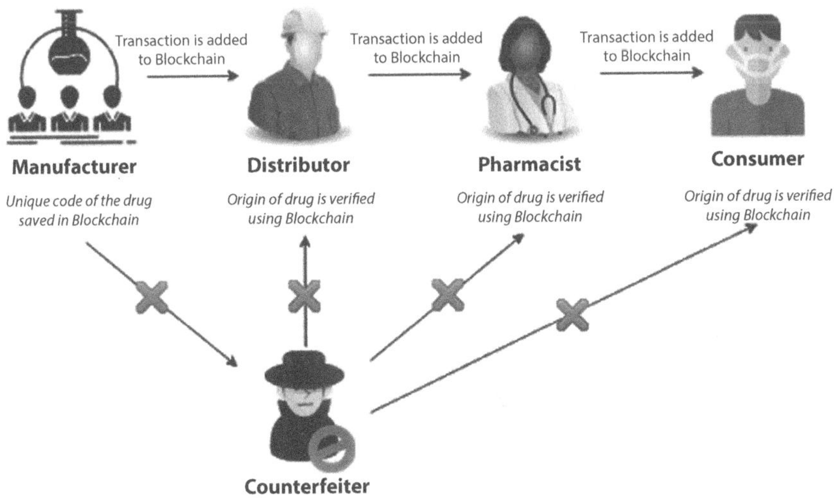

Figure 8.11 Blockchain integrated SCM system for countering drug counterfeiting.

8.4.6 Clinical Trials

Data generated during Clinical Trials are used for research purposes or approval of new treatments. The integrity of this data is essential because it holds importance for various stakeholders like the researchers, journal editors and publishers, drug and device companies, government regulators, and essentially, prospective patients and the general public. Earning of the companies that have invested in its research and development is at stake. Thus, the security of the data obtained from clinical trials is of utmost importance [20].

There are several risks to Clinical trial data like alterations, by mistake or even with malicious intent to achieve the expected result and false or selective reporting of an outcome of an experiment. Such manipulation of data results in variation from actual bona fide outcomes, thus causing problems for the stakeholders and even serious health hazards for the prospective patients.

The introduction of Blockchain in Healthcare for improved Clinical Trial Dataset Management would not only provide privacy to patients' records but also ensure that the data shared amongst the researchers and publishers did not tamper. This would maintain confidence between the various stakeholders as now, only authorized and trusted parties are given access to its creation and manipulation.

8.4.7 Public Health Management

The health of the general public can be equated to prosperity and growth in the country. The government is mainly involved in Public Health

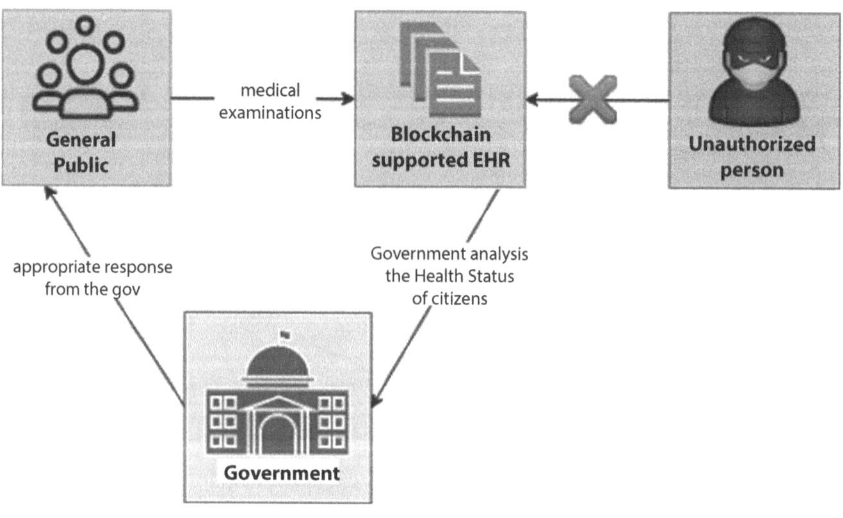

Figure 8.12 Blockchain technology in public health management.

Management because it is elected to power for its promises. A few of them being policies and reforms regarding Public Health Management will be implemented along with other policies. Keeping records of individual citizens like the Electronic Health Record (EHR) and analyzing it would help the government identify if the population is faring well or not. Figure 8.12 depicts how EHR can prevent unauthorized access.

8.5 Challenges of Blockchain in Healthcare

Already discussed are the challenges faced in implementing blockchain technology. Now we are going to look at the challenges specific to implementing it in Healthcare.

Storage: With rapid digitization, now facing storage issues because of the generation of tremendous amounts of medical records, images, documents and laboratory reports and data collected by sensors and wearable IoMT devices. Blockchain architecture is limited by on-chain data storage, meanwhile, our current system produces a lot of data and yet to find a solution to this problem.

Costly: Blockchain architecture is very costly to implement and it only becomes even more expensive as it increases the scalability of our system. Its implementation requires the most effective infrastructure and interconnectivity, which startups cannot afford easily. Data storage and current consensus algorithms are contributing to its expensive nature. Addressing the cost-effectiveness of the technology should involve addressing data problems and adopting alternate state of the art consensus algorithms which are concerned fields of study revolving around this technology.

Modification: The immutability feature of blockchain doesn't allow any modification. Blockchain applications should be developed keeping in mind the immutable nature of technology such that changes are only committed to the chain once everything is confirmed.

Scalability: The scalability issue needs to be taken seriously to ensure blockchain popularity. Because of its decentralized nature, scalability is somewhat easier, however, there are private clinics, small healthcare centers, small organizations, etc. which might not have the architecture needed to support the system. Blockchain technology also requires tremendous computation power, which calls for higher electricity consumption.

Privacy and regulations: Blockchain maximizes the security of its content repeatedly. Cryptographic, decentralized and immutable architecture makes sure that its content is at the highest security. All the large medical information files of the patients are sensitive information and therefore, it is often risky to store a copy of these data in every node. The foremost demanding problem for the blockchain technology currently being used is the storage of Electronic Health Record (EHR) and Personally Identifiable Information (PII).

Lack of Knowledge: There is a lot of reluctance in regards to adopting this new technology from various organizations that play a key role in healthcare systems. These organizations are unaware of its working and benefits. They must educate themselves about this rising technology. Blockchain is still evolving and a lot of organizations are not aware of its existence. There's still a lot of work needed to promote it across various domains.

8.6 Future Research Directions

Blockchain is still in its evolving stage and its implementers need to take appropriate steps for its development in the right direction. This is crucial to bring in efficiency and reduce latency whilst not compromising the security, required to assist the healthcare infrastructure.

Mention a few challenges to the implementation of Blockchain into Healthcare and provide appropriate solutions which should be the primary aim of researchers and implementers of Blockchain specific to Healthcare Infrastructure.

Table 8.1 Future directions.

Future research directions	Description
Green Blockchain	It was estimated that bitcoin consumed more energy than the entire country of Switzerland. With more emphasis on reducing carbon footprint, more energy-efficient methods are needed to implement blockchain if want to be a part of the zero-carbon emission future.
Technology Adoption	There is a need to educate and create awareness about Blockchain and its practical usage and knowledge among various stakeholders. Only then, it would be possible to implement Blockchain across various sectors of healthcare and have a unified, efficient, automated healthcare system.

(Continued)

Table 8.1 Future directions. (*Continued*)

Future research directions	Description
Technology Advancement	• Currently, blockchain majorly uses a proof-of-work consensus algorithm. It vastly increases the time taken to mine new blocks and verify current transactions. Smarter algorithms and increased computing power are needed so that they can be implemented more easily, efficiently and with less cost for implementation. • Huge computing power is needed to mine a new block, only organizations with immense computing powers are currently mining new blocks. This contradicts the idea of having a 100% distributed system eliminating 3rd parties for authentication since big organizations are allowing us to use this technology. This technology needed for mining requires powerful computers that are well out of a common man's reach.
Feasibility	There is a need to realize that blockchain might not be the only solution when met with an obstacle. Its cost-effectiveness and time taken to implement could be a decisive factor for its feasibility. That is why there is a need for a shift from the Ideation to Prototype development and testing stage to check the idea's feasibility.
Scalability	Scaling up Blockchain is difficult. With its growth, the need for computation power also increases. More research work is needed to effectively and efficiently scale up our blockchain while also keeping in mind its adaptability with future needs.
Interoperability	As there would be applications developed on various platforms, based on Blockchain in Healthcare, there is a need for these applications to function together properly. Thus, integration of the apps on different platforms will help achieve interoperability and facilitate an end-to-end integrated Blockchain network for Healthcare systems [1].

8.7 Conclusion

Since Blockchain's invention in 2008 by Satoshi Nakamoto for bitcoin, it has inspired various other applications based on Blockchain. Although still in the developing stage and constantly improving, its implementation into the Healthcare infrastructure could be the key to a quick, tamper-free, automated and healthy future. This chapter started the discussion with an overview of healthcare and the issues and challenges it faces. It then explored the underlying technology for implementation, Blockchain, its working, applications and the drawbacks. Further, it discussed the implementation of Blockchain into the Healthcare system followed by present challenges that discourage organizations from adopting this technology.

There is potential in this technology, now started to see the inception of Dapps (Decentralized apps). The future seems promising with technology and more people will start to adopt it with the evolution of technology.

References

1. Krishnan, S., Balas, V.E., Golden, J., Robinson, Y.H., Balaji, S., Kumar, R. (Eds.), *Handbook of Research on Blockchain Technology*, Academic Press, Elsevier, USA, 2020.
2. Health IT Government Website USA, 2020, https://www.healthit.gov/topic/health-it-and-health-information-exchange-basics/health-it-and-health-information-exchange.
3. Lakshminarayanan, S., Role of government in public health: Current scenario in India and future scope. *J. Family Community Med.*, *18*, 1, 26, 2011.
4. Sravan, N.P.V., Baruah, P.K., Mudigonda, S.S., Use of Blockchain Technology in integrating Heath Insurance Company and Hospital . *Int. J. Sci. Eng. Res.*, 9, 10, 1–6, 2018.
5. Rosic, A., *What is Blockchain Technology? A Step-by-Step Guide For Beginners*, December 2020, *https://blockgeeks.com/guides/what-is-blockchain-technology*.
6. Christensson, P., "P2P Definition." TechTerms, Sharpened Productions, 2006. Web. 30 December 2020. https://techterms.com/definition/p2p.
7. https://builtin.com/blockchain
8. Zheng, Z., Xie, S., Dai, H., Chen, X., Wang, H., An overview of blockchain technology: Architecture, consensus, and future trends, in: *2017 IEEE International Congress on Big Data (BigData Congress)*, 2017, June, IEEE, pp. 557–564.
9. Nick, A. and Hoenig, L., Consensus Mechanisms in Blockchain Technology, 2020, https://www.lexology.com/library/detail.aspx?g=e30e7d54-3c7f-4ca0-8a22-478227a9b5ec.
10. Nguyen, G.T. and Kim, K., A survey about consensus algorithms used in blockchain. *J. Inf. Process. Syst.*, *14*, 1, 101–128, 2018.
11. Alharby, M. and Van Moorsel, A., Blockchain-based smart contracts: A systematic mapping study. *Fourth Int. Conf. Comput. Sci. Inf. Technol. (CSIT-2017), arXiv preprint arXiv:1710.06372*, 125–140, 2017.

12. Iredale, G., Top Disadvantages of Blockchain Technology, November 2020, https://101blockchains.com/disadvantages-of-blockchain/.
13. Golosova, J. and Romanovs, A., The advantages and disadvantages of the blockchain technology, in: *2018 IEEE 6th Workshop on Advances in Information, Electronic and Electrical Engineering (AIEEE)*, pp. 1–6, IEEE, 2018, November.
14. Anupam, S., What are the Major Limitations, Challenges in Blockchain, December 2020, https://inc42.com/features/what-are-the-major-limitations-challenges-in-blockchain/.
15. Katuwal, G.J., Pandey, S., Hennessey, M., Lamichhane, B., Applications of blockchain in healthcare: Current landscape & challenges. *arXiv preprint 2018arXiv:1812.02776K*, December 2018.
16. https://www.aeris.com/news/post/the-internet-of-medical-things-iomt-and-the-new-patient-journey-how-iot-is-helping-to-navigate-healthcare-challenges/#:~:text=IoMT%20networks%20can%20use%20this,checkup%20as%20soon%20as%20possible.
17. Shroff, R., *Blockchain in insurance: Use cases and implementations*, 2019. https://medium.com/swlh/blockchain-in-insurance-use-cases-and-implementations-a42a00ebcd91
18. https://www.cbinsights.com/research/blockchain-insurance-disruption
19. Haq, I. and Esuka, O.M., Blockchain technology in pharmaceutical industry to prevent counterfeit drugs. *Int. J. Comput. Appl.*, *180*, 25, 8–12, 2018.
20. Maslove, D.M., Klein, J., Brohman, K., Martin, P., Using blockchain technology to manage clinical trials data: A proof-of-concept study. *JMIR Med. Inf.*, 6, 4, e11949, 2018.
21. Phillips, M., Benefits and Challenges of Blockchain Technology in Healthcare, 2018, https://seeromega.com/10-benefits-challenges-blockchain-technology-healthcare/.
22. Rudra, B., Impact of Blockchain for internet of Things Security, in: *Cryptocurrencies and Blockchain Technology Applications*, pp. 99–127, 2020.
23. Sharma, D.K., Kaushik, A.K., Goel, A., Bhargava, S., Internet of things and blockchain: integration, need, challenges, applications, and future scope, in: *Handbook of Research on Blockchain Technology*, pp. 271–294, Academic Press (Elsevier), USA, 2020.

Section 3

CASE STUDIES OF HEALTHCARE

9

Organ Trafficking on the Dark Web—The Data Security and Privacy Concern in Healthcare Systems

Romil Rawat[1]*, Bhagwati Garg[2], Vinod Mahor[3], Shrikant Telang[4], Kiran Pachlasiya[5] and Mukesh Chouhan[6]

[1]*Department of Computer Science Engineering, Shri Vaishnav Vidyapeeth Vishwavidyalaya, Indore, India*
[2]*Union Bank of India, Gwalior, India*
[3]*Department of Computer Science and Engineering, IPS College of Technology and Management, Gwalior, India*
[4]*Department of Information Technology, Shri Vaishnav Vidyapeeth Vishwavidyalaya, Indore, India*
[5]*Department of Computer Science and Engineering, NRI Institute of Science and Technology, Bhopal, India*
[6]*Department of Computer Science and Engineering, Government Polytechnic College, Sheopur (MP), India*

Abstract

On the dark web (DW), criminal's are technologically knowledgeable so security agencies must keep up to speed with all data sources in endeavors to find criminals and identify their movement. It has gotten progressively harder to follow this area and monitor criminals and terrorists, as they find and grow new approaches underground for their online activity. Due to the shortage and trouble of safeguarding transplantable organs, patients should normally look out for extensive harvesting for a long time to get a coordinating kidney and crucial organs. This restricted accessibility has made an unlawful commercial center for well-off beneficiaries to try not too considerable delay times. Agents organize such organ transplants and gather the greater part of the installment utilizing cryptocurrency that is now and then directed to finance other criminal operations. To gather and dispense installments at abroad areas, they generally resort to money laundering for finding the money at worldwide spots without being seen by security agencies.

**Corresponding author*: rawat.romil@gmail.com

Kavita Sharma, Yogita Gigras, Vishnu Sharma, D. Jude Hemanth and Ramesh Chandra Poonia (eds.) Internet of Healthcare Things: Machine Learning for Security and Privacy, (191–216) © 2022 Scrivener Publishing LLC

With the ease Internet getting to show up in a portion of the influenced nations, online activities of organ trading at the hidden world of DW are utilized to wrongfully exchange human vital body parts.

Due to the strictness and limitation of storing and preserving vital organs, the patients always require fresh donors of organs, which could be transplanted to needy people. This waiting time sometimes proves to be life-threatening for patients. To overcome the disastrous situations, the illegal techniques are called using agents who coordinate among donor and receiver by the exchange of money, but at a secret place, because the trade is illegal in most of the countries. So, it's like a covert mission to take both the persons, where safe operation is done, doctor, who is also part of illegal organ trade. Nonetheless, interviewees of one investigation expressed that even though purchasers pay more, vendors get far less for their organs. Also, a few beneficiaries will in general become intermediaries at a later stage. Numerous cases examined illegal trading of organ exchanges have found, the dealers crumble after concealed deals and are exposed to social disgrace so they at last lament their deal in 94% of the cases. This Propose work presents a building structure to evaluate the danger of the human organ trafficking trade at the online social stage and to distinguish online presence and channels of correspondence medium.

Keywords: Organ trafficking, dark web, cyber crime, data security, HealthCare System

9.1 Introduction

Consistent expansion in the number of online attacks to syndicates and nation-states leaves them impotent and at significant risk of protection and data privacy [1]. About irrefutable web-generated vulnerabilities and web crimes linked to the digitization of society, the predictions of the fourth mechanical revolution accompany an immense expense to bear. The usage of an elective Internet correspondence stage, for example, DW is considered to be a safe house for planning and executing various types of web crimes. The correspondence stage that conceals Internet convention locations of people will forestall the assurance of sources and objections of web-generated-Vulnerabilities [1]. People getting to the Internet are adjusted to a normal online world related to online social platforms without the full consciousness of conceivable web-generated-Vulnerabilities presented by unlawful utilization of DW. The concealed web path is also called a Concealed web path and Organ Trafficking is also called organ harvesting.

Human harvesting is a horrifying unspeakable atrocity that is annihilating the lives of an expected million individuals in the world today. Human harvesting still lives and grows at an alarming pace among us. Truth be told, today there are a larger number of people in chains than human harvesting left us at the time. In the structures of our cutting edge culture, there are numerous causes of growth that have contributed to the development of this curse against mankind. The ascent of the internet and especially the DW are empowering traffickers to work their misconduct without any difficulty in this day and age.

International Trade Merchandise Tracking Agencies define the harvesting of persons as the recruiting, transportation, travel, preservation or receipt of persons, using danger or use of authority or different ways of coercion, abduction, misrepresentation, repetition, abuse [2] of strength or position of weakness or installation or reception [2]. Abuse should require, at least, the exploitation of other people's prostitution or various ways of sexual abuse, limited employment or facilities, bondage, or activities such as subjugation, subjugation, or organ expulsion.

Human harvesting is a criminal action that, always crossing national lines and institutional reach, is regularly highly orchestrated in nature. Traffickers conduct this lucrative business, using the latest technologies available to hide their illegal activities with little fear of being obtained. The developments in creativity are quickly offering them more spaces to stow away.

In the Internet convention stack, DW represents a layer beneath the deep web that cannot be listed by universal web crawlers. Limiting traditional web indexes to surface web search has made DW a safe house for all forms of web-generated vulnerabilities [2] to be engineered and executed. The surface web is inherently available by universal web crawlers that can look about 5% of the whole web [4], however, DW, which isn't open by customary web indexes requires particular programming, for example, the invisible internet project (I2P) to be gotten to [2]. Black web has gotten critical press inclusion nationally and internationally as huge bootleg merchandise for engineering diverse Cybersecurity [46] web peril Wright [3]. It contains questionable materials, for example, fake money, travel papers, and personality harvesting. It has become a shelter for the deals of dangerous weapons, unlawful medications, and spilt data and youngster's sexual entertainment.

Structure of the Chapter

The rest of the chapter is organized as follows. Section 9.2 covers inclination for Cybersecurity web peril, Section 9.3 covers literature review, Section 9.4 shows about the market paucity or organ donors, Section 9.5 covers organ harvesting and transplant tourism revenue, Section 9.6 shows about social web net crimes, Section 9.7 covers dark web—the frontier of illicit human harvesting, Section 9.8 covers organ harvesting apprehension, Section 9.9 covers result and discussions and finally, Section 9.10 concludes this chapter.

9.2 Inclination for Cybersecurity Web Peril

Patients can watch for transplant harvesting for a long time due to the lack of transplantable organs (sometimes six years for kidney transplants) and others kick the bucket at transplant hold-up periods. To prevent this substantial wait, rich patients in nations that lawfully or unlawfully motivate such transplants buy organs from poor persons. A few institutions and centers do not place legitimate limits on such transplants, fostering a clandestine commercial hub for illicit organ trafficking in this manner. Given the discoveries from the dissection of the collected data on organ procurement, the cost of kidneys has risen significantly somewhere in the 2008 and 2015 range, and the cost of livers has also increased by an outright amount equal to that of kidneys. These higher benefits draw in more unlawful organ harvesting far and wide. Nonetheless, interviewees of one investigation expressed that even though purchasers pay more, dealers get far less for their organs. Also, a few beneficiaries will generally become representatives at a later stage. Numerous researchers who examined concealed organ exchanges have found that the health states of dealers break down after deals and are exposed to social shame so they in the end lament their deal in 94% of the cases [5]. The locales' distributers on the black Web are unknown and covered up. Clients are gotten to on the DW to impart data to little hazard and to be undetected (unknown). The entrance of clients namelessly is basic for the DW, which as of late is upheld by the codification tunnelling for monitoring security. The MR manages the Secret Web material. It is an unknown web network and the anonymous software has to access it. In 2002, the project was dispatched by the US Naval Research Laboratory [6] to allow mysterious online correspondence. I2P is another web network with congestion in its borders on the Internet and is used for unexplained communications, traffic codification of users, and so on. It has a better and more unwavering website quality [1, 5]. Figure 9.1 below shows the architecture of the DW portion in the web environment.

The probability of customer position is poor in the Masked Network, and this has facilitated the development of legal and illegal activities among the congestion of this section of the Internet [2]. Using unknown and dispersed hubs with such web-net gatherings, a masked network can be done. Hidden routing has the product name that introduces on the system web net that mind and deal with its associations. It empowers clients to get to the web forum through virtual passages where individuals and syndicates can disseminate data through open web nets without trading off their privacy. MR empowers clients to course their congestion through clients'

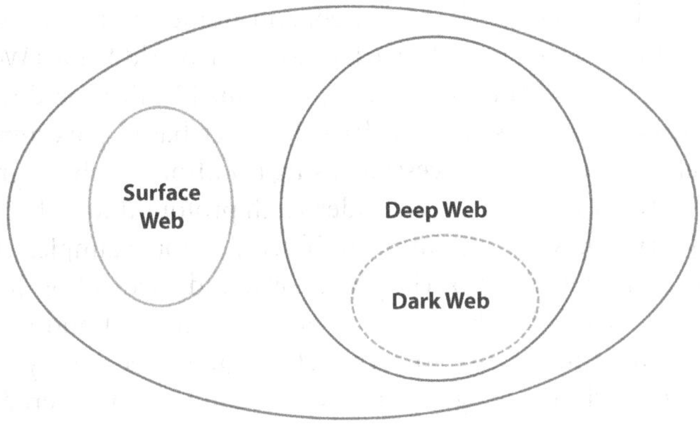

Figure 9.1 Three Layers of web.

systems so altogether that traffic isn't followed back to the beginning clients and cover their personality. To pass the data starting with one layer then onto the next layer, concealed web path has made transfers on systems that help data through its passages everywhere on the world [2].

A. Cybersecurity Web Peril on Hidden Web

There are several ways to receive information from the Internet freely, including a little bit of this information. Different layers lie under the Internet and are expected to be even wider than the surface network. For example, Google, the specific basic programs that individuals are acclimatized with, is used to get to the Internet content, but DW is a type of Internet substance that is not ordered by this web crawler definition for various specialized purposes. Thusly, these substances would not be naturally open through a standard web crawler [14]. Besides, concealing Hidden content was deliberate, and the capacity to share data while staying mysterious is one of the cardinal reasons that clients are getting to DW [6]. Scarcely any examinations that dissected darknet drug commercial centers have learned various kinds of substance situated on an onion ecosystem and played out an investigation of venders of the drug merchandise [9], decreased a bunch of transfers to monitor and recognized wellsprings of unknown hidden web net congestion. Specifically, concentrates by [7, 8] uncovered that endeavors have been made to diminish the number of transfers to be looked and they investigated different techniques to distinguish mysterious clients in web net associations using routers and transfers. International Trade Merchandise investigation agencies [3] represent everyday domains are taken off day by day, which establishes new domains being made day by day. At this point, no study has produced a sign of the true scale of the surface web, deep web, and hidden web network of the deep web or hidden web net. The surface web is a World Wide (Web) compartment accessible by traditional web programs [2]. Profound web refers to properties and records, such as legal Internet-based government harvesting that is not explicitly accessible using traditional web programs and hyperlinks. DW is a compartment underneath profound web that is widely utilized by criminals for executing web crimes, for example, youngster erotic entertainment, deals of risky weapons, and medication harvesting [4]. DW is an encoded web net that uses public Internet and is purposefully shrouded utilizing shared web net to guarantee anonymity [28]. The survey of various web-generated-Vulnerabilities engineered on hidden web net is considered in this segment for the equivalent of clarity.

B. Healthcare Data Security Threat

The healthcare suppliers utilize electronic harvesting (e-harvesting) for patient monitoring and it contains a few factors related to health data systems. If one patient has alluded to the next clinic the harvesting will effectively move to the current healthcare supplier and in this way, the health supplier can follow all the past health-related boundaries. This data is routinely kept up under and by recognizable data and ensured health data. All the services brought out through by e-Prescribing programming and monetary [10] data through medical charging programming. Cyber illegal activity is unmistakable in the healthcare trade because the outcomes are identified with patient health and wellbeing. Hence, when health data is taken the patient can't adjust his fixed data (Date of Birth, blood classification, and so forth) and this can be utilized for a wide scope of violations. The managing technique in hospitals can contrast essentially from patient to patient and is confounded to computationally display, this frequently requires transparency and consequently shaky codes which prompts data-web peril. Antivirus programming is huge with significant highlights, for example, break tests, control of actual access, and the support of often rebuilt reinforcements in which data can be stored disconnected.

Progressions, for example, the IoT (Internet of Things) [5, 46] give the highlights of distant medical [11] consideration and carefulness in healthcare delivery. However, clinical services and wellbeing are offset with privacy. Gadgets are exceptionally organized in the medical clinic system and colossal measures of data are gathered through the medical facility and a wide range of hospitals. They regularly don't have the fitting security methods because the underlying assets to proficiently utilize security estimates, the codification danger demonstrating exercises, and malware discovery. The principle of these gadgets to work under conditions are not reasonable and territory with a frail scope of the website. Additionally, the system has high battery life, and the working programming ought not to outdate rapidly. Leaders should assess the anticipated life expectancy of gadgets before procurement. In the mix, gear assurance is essential to medical-gadget security. Hospitals and producers, with help from confirming specialists, ought to build up a fixing strategy that limits gear personal time and empowers ideal updates through a coordinated effort with the outer assembling web net and interior partners.

9.3 Literature Review

Organ harvesting, a rewarding illicit global trade, is a less commonly explored method of human harvesting by anti-human harvesting partners due to its perplexing and often clandestine nature. Among founders of public policy and general mindfulness campaigns [6], harvesting sex and/or work are the most widely regarded forms of human harvesting [7]. In either case, because of popularity and relatively low rates of law enforcement, organ harvesting holds a basic position for transnational organized misconduct bunches. In the shadows, organ harvesters [12] benefit, while the solitary thing that is sensed is their dangerous medical impression. It leaves weak populaces, otherwise known as contributors, and first world recipients, otherwise known as beneficiaries, open to serious abuse and a long period of health consequences. This type of illegal exchange [13] additionally leaves the private sector, specifically the monetary business, vulnerable to being an unconscious course for its assistance. Although, with the correct preparation and bringing issues to light, monetary foundations may assume a vital part in exposing organ merchants via the monetary path they give up.

Human following here and there alluded to as present-day bondage in the 21st century keeps on being a test to the international website. Numerous researchers and syndicate concentrate their endeavors to diminish human following far and wide. The creator clarified that an in-wrinkled remarkable quality in relative hardship can lead people to look for additional dangers, in this manner oppressing themselves and their kids at more danger of being misused [14, 15]. Examined criminal possess and their trade, as usual, utilizing proof-based research. They show that different factors, for example, financial and social powers can likewise in-wrinkle expected casualties' weakness. Even though addressing the issue beforehand is better than addressing any aftermath later, the thorough examinations in organ following are a long way behind as referenced in [17]. Take such a test by setting up an organ hazard model that partners with the assessment of the damage measured. This threat model is intended to illustrate genuine organ behaviors to discern the possible organ trackers that can help intercede or even forestall the accompanying organ. Organ dealing casualties [47], likewise with most illegal exploitation casualties, are by and large defenseless populaces [48]. There are uncommon occasions where casualties are put under sedative and wake to discover their organs missing or are killed for their organs. In general, the pressure of living givers is more normal (Future for Advanced Research and Studies, 2016; United Nations, 2011). It is generally regular for casualties of organ dealing to be selected through specialists, who are people who enroll organ providers and interface them with organ beneficiaries (Kelly, 2013; United Nations, 2011; United Nations, 2015). Spotters/agents are typically individuals who might be from similar networks or the identity of a weak populace to assemble confiding seeing someone simpler. The spotters at that point make vows to the organ providers like enormous amounts of cash or delivery from obligation, and persuade them that the organ isn't required. Explicitly on account of kidneys, the most usually gathered organ from living contributors, enrollment specialists will tell casualties that the kidney will develop back, having two kidneys is unnatural, or that they have a huge and a little kidney and expulsion of the little kidney is innocuous (United Nations, 2011; United Nations, 2015) [48]. Casualties seldom get everything of cash guaranteed, if they get any remuneration whatsoever (United Nations, 2011; United Nations, 2015). Sometimes, the post-evacuation medical care costs professionally organ dealing casualty add to their past obligation and deteriorate their monetary circumstance (Kelly, 2013).

China is particularly notable for organ dealing, with reports of detainment facilities executing detainees to wrongfully reap their organs against their assent when they are a possible counterpart for a beneficiary (Paul, Caplan, Shapiro, Els, Allison, and Li, 2017). The United States, Canada, the Czech Republic, and Israel have started to make a move to keep their residents from participating in transferring the travel industry including China (Doctors against Forced Organ Harvesting (DAFOH), 2019). On the other hand, organ acquisition through buy-in Iran is now and then lawful, as Iran is the lone country on the planet where it is lawful to sell organs (Paul, Caplan, Shapiro, Els, Allison, and Li, 2017). Iran has set up a base cost for organs at $4,600, yet that is just when the organs are acquired legitimately which is frequently not the situation, as needy individuals experience handles and are paid an obscure, under-the-table value (Krishnan, 2018). One would trust that a couple of disengaged nations are significant members in organ dealing, however lamentably it is a worldwide business where organs are purchased from poor people and offered to the rich (European Union, 2015). Organ beneficiaries partaking in transfer the travel industry are generally going from the United States, Canada, Saudi Arabia, the Gulf states [49], Japan, and Taiwan into Costa Rica, Panama, Ecuador, Colombia, Egypt, Kosovo, Cyprus, Israel, Azerbaijan, China, and the Philippines to get organs (European Union, 2015). While not hot areas for relocating the travel industry, the accompanying nations have been distinguished as organ beneficiaries.

9.4 Market Paucity or Organ Donors

When depicting organ trafficking, there is sometimes doubt as to how this wrongdoing might occur. Worldwide Financial Transparency tests that 10% of all organ transplants, including lungs, heart, and liver, are conducted by trafficked organs. Moreover, kidneys [19] are the most visible organs that are unlawfully traded, evaluating the Vital Organs (kidneys) [13] are internationally exchanged on the bootleg products.

When contrasted with normal sit tight occasions for organs in created nations, one can begin to all the more likely understand the need to be redirected to underground web merchandises. It is assessed that the normal hang tight an ideal opportunity for a kidney is 4 years for certain holding up to 7 years [18]. At the International Trade Merchandise, the normal hang tight an ideal opportunity for a kidney is 4 years as per the National Kidney Research Foundation [4].

A. Covert Organ Exchange Operations

Where harvested, trafficked organs can be transplanted to recipients in the most trustworthy of hospitals in large metropolitan areas around the world, but shoddy workplaces in houses have also been the forgotten areas for such transplants. Organ harvesters often coordinate the benefactor's enlistment from a position of vulnerability, and victims are not properly tested on their capacity to be healthy giver. Frantic patients wanting an organ may fall victim to a trafficker who could act as a legal representative of an organizing agency of an unselfish organ. On both sides of this situation, monetary exploitation has a vital consequence. Also, organ smugglers may be active in various forms of human harvestings, such as the harvesting of sex and/or work [17]. Cases occur when, as a perpetrator of organ trafficking, an organ giver may have been a victim of sex harvesting and/or job harvesting, allowing a phased state of abuse. Any country in which organ procurement operations are carried out is seen in Figure 9.2 below.

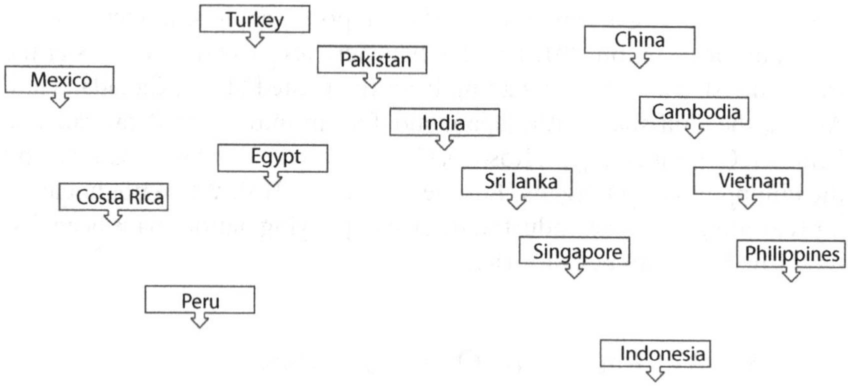

Figure 9.2 Globally organ transplantation activities.

B. Expanding the Human Harvesting Lexicon

Reaping in individuals involves choosing, transferring, retaining, or receiving individuals, tactics for pressuring or using strength or various forms of pressing reasons, trapping, coercion [21], uncertainty, misuse of power or place of vulnerability, or granting or tolerating portions [21] or advantageous conditions for obtaining consent and obtaining consent. At either rate, abuse can include the abuse of other people's trafficking or other forms of sexual abuse, forced labor or administrations, servitude, or activities such as coercion, enslavement, or organ departure [15]. Figure 9.3 below shows the organs used for selection.

The buying and exchange of organs are prohibited in many nations (e.g., Iran is the lone country in the world where it is lawful to buy and trade an organ, although this particular situation only applies to its inhabitants) [11, 23]. Alternatively, there are not many laws that prohibit a person from leaving one's country to procure an organ from someone abroad. To be told, various syndicates oblige the travel industry to transplant but suggest that recipients are only coordinated with willing donors. It is impossible to know exactly how much transplantation the travel industry generates worldwide per year, but it is known that the illegal trade of organs creates billions of companies annually [16]. It is difficult to detect associated currency flows for both law enforcement officials and anti-money laundering experts. This is attributed to a host of reasons, such as the absence of homegrown laws banning citizens from moving internationally, the transnational definition of misconduct, and the services of suppliers who understand the laws identified with organ harvesting all around okay to avoid them via shell syndicate and distilled (legal) contributions through an open web forum.

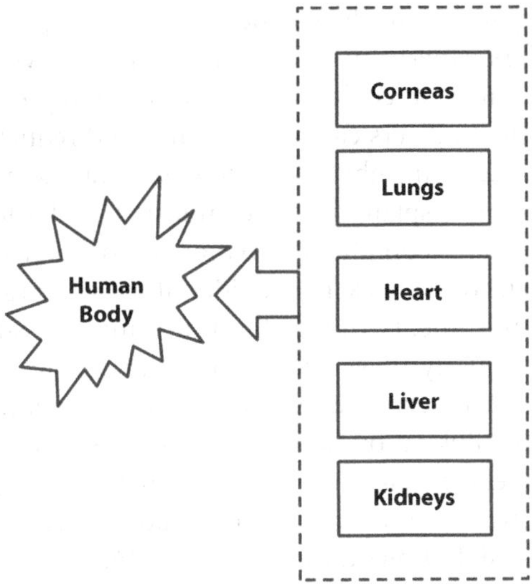

Figure 9.3 Human body organ parts.

C. Illicit Currency Concealment

While it could be hard for banks to distinguish monetary exchanges identified with organ harvesting, it isn't unthinkable as there are a few indicators accessible. These warnings could incorporate the accompanying indicators and might be harmless all alone however when consolidated, might introduce dubious conduct:

- Wire moves to substances in high-hazard wards with names that incorporate a variety of medical.
- Methods of installment, for example, wires installment, email money move, and mass money withdrawal.
- Payments among noble cause and medical the travel industry destinations.
- Credit card installments to travel services, aircraft, or lodgings, before the development of money and travel.
- Medical the travel industry web-forum that offer transplant activities abroad that suggest using their own confided in homegrown doctors preceding voyaging.

One thing to note is that although traveling overseas to get an organ in individual countries could be legal, related monetary exchanges will, in any case, be considered as reportable in various ways as the demonstration of purchasing an organ within their nation of citizenship may be unlawful. This position offers experts an interesting point of view that far exceeds that of law enforcement since they are in a position to present information that law enforcement officers cannot recognize and require it. Knowledge assembled by monetary insight units inside monetary syndicate related to organ harvesting or transplant the travel industry can be additionally scattered to international accomplices by national monetary insight units. As recently expressed, covering exchanges identified with organ harvesting is no simple accomplishment. This method of bringing issues to light may end up being a similarly powerful device in hindering organ harvesting while at the same time expanding insightful information toward revealing exchanges.One illustration of how mindfulness is being raised, inside the setting of monetary insight units and organ harvesting, is through the Project Protect activity in the International Trade Merchandise, dispatched by the monetary insight unit's master [22] in 2016.

D. International Trade Merchandise

When obtaining their national human harvest rules, the Foreign Trade Merchandise omitted organ harvesting as a method of human harvesting. Since the Palermo Protocol [41], the discussion of organ procurement has consistently gained a public approach. A gathering of key partners in the worldwide fight against organ harvesting met in 2008 to form the Istanbul Declaration, which, after Istanbul, developed pivotal new definitions for organ harvesting and transplantation in the travel industry and created promising practices for organ exchange management. The arrest of organ offenders in web-based Foreign Exchange Merchandise locations has been sparked by informing reputable specialists. In 2014, the International Security Council adopted a Convention against the Harvesting of Human Organs, which became applicable as of late January 2018 [17]. This is a profound enhancement with an all the more generally limitless supply of organ procurement as the key legal part. The degree of organ procurement is as yet uncertain as to the number of such transplants conducted each year as of today. In comparison, the total convergence of the problem within the area of human harvesting is still missing in everything. It is imperative to engage in a sustainable public–private syndicate to combat organ harvesting between various forms of transnational coordinated misconduct.

E. Overseas Pervasiveness

Data on the pervasiveness of organ harvesting is limited, and the study of the travel industry's organ harvesting and transplantation as transnational problems has been largely based on empirical analysis. Because of the absence of standardization of data announcement across jurisdictions, data obtained by explicit wards reduced an opportunity for transnational review and reaction. Accessible statistics on harvesting in people for the evacuation of organs are inadequate in those concerned with harvesting in people for the expulsion of organs (counting casualties) provide no impetus to target [healthcare] providers [25], who end up treating people who have acquired organs abroad might be prohibited from supplying details to specialists due to patient privacy issues, Also, definitional problems and disarray lead to powerless reporting and review [26].

9.5 Organ Harvesting and Transplant Tourism Revenue

Broad gauges have been made about the scale of the unlawful trade of human organ transplants, irrespective of the impediments to usable records. Human rights misconduct, as classified in various foreign treaties, is the harvesting of human organs and the harvesting of persons for tissue evacuation. The trade-in human organs are closely denied all over. Despite these bans and restrictions, the study alert organization [27] has assessed the illegal industrial trade of human organs. Figure 9.4 provides a gauge of paces of worldwide trade transplantation and costs paid by beneficiaries as revealed in the Global Financial Integrity Study distributed in 2017.

The comparative costs for commercial transplants have been announced in numerous reviews. In 2006, for example, a report by the World Health Organization showed that the expense of renal transplantation went between anywhere in the region of $75,000 and $160,000 [28]. Deep transnational criminal web networks, involving predatory intermediaries, human traffickers, and incompetent doctors, empower this illicit trade.

A. Feature of the Hidden Web

The significant segments of the DW are browsers to get to the codification procedure to scramble the data, Virtual Private Webnets for communicating

S.N.	Organ	Global illicit transplants	Price range
1	Kidney	7,995	$50,000-$120,000
2	Liver	2,615	$99,000-$145,000
3	Heart	654	$130,000-$290,000
4	Lung	469	$150,000-$290,000
5	Pancreas	233	$110,000-$140,000

Figure 9.4 Overview of the organ market.

the data, and steering calculation [31]. Program isn't sufficient to remain unknown yet also, you need to utilize a decent Virtual Private Webnet.

The TOR program utilizes multiple levels of intricate codification and random steering is used to ensure your personality, but the dilemma is always there as the observer will even now peruse the messages you send or receive. For instance, uprightness, clarification, privacy, and non-disavowal were intended to provide security perspectives. The PGP [20, 24] is inherently dependent on unequal codification. Two separate keys are used in codification, such as Public Key and Private Key, to encrypt and unscramble the records. PGP (Pretty Good Privacy) can be used for authentication purposes as well. PGP functions in an alternative manner for confirmation. It utilizes a combination of hashing and codification of the public key. It utilizes a combination of mystery key codification and public key codification to provide anonymity. Therefore, computerized marks use one mystery key, one hash power, and two public-private key sets. The use of PGP Codification has several benefits. Initially, because it cannot be seen or taken by someone on the Internet, the data is constantly assured. Data or information can be safely exchanged over the Internet. Whenever they are lost, erased texts or other touchy data will not be restored. Also, attackers [29, 33] can't taint the message or messages. This form of codification confirms the sender's data with the intention of not being captured by the outsider. It is all but impossible to use. By using symmetric square codification, classification is given. The instrument for authentication [38] is provided by specialized signatures [43].

Several projects have been built to get to the DW. The concealed path is the most widely used application for DW. During the 1990s, it was developed at the US Lab by Paul Syverson, Michael G. David, and Reed Goldschlag [32]. In C, Python, and Rust, the Concealed Route was written. In September-2002, the alpha form of the concealed route was dispatched. At the onion directing strategy [12], it chips away. The client data is initially scrambled in this technique and then data is transferred by multiple transfers (transitional systems) available in the site network.

9.6 Social Web Net Crimes

The well-known press inclusion of the DW investigated above doesn't talk about Social Webnet Analysis (SNA). SNAs, it shows up, are outside the media philosophy of the DW. For writers covering the Hidden web, the possibility of DW clients taking part in Social Webnet examination for Social Webnet investigation for criminal, in the nonexclusive structures presently perceive: For instance, friending, preferring, following, posting: is by all accounts incredible, That is, there is no verbalization of with either unlawful exercises, liberal free discourse on the DW is presently connected with the utilization of genuine world-personalities, is viewed as a protected and satisfactory online practice. Facebook explicitly has been very effective in cultivating a culture of genuine characters, however, it did as such even with a culture of anonymity [36].

The Social Webnet investigation for Social Webnet examination for the criminal is more fragrant of social web net before Facebook, that is, social web net with nom de plumes. Truth be told, Facebook's forceful planning of certifiable characters is a contributor to the issue that the Social Webnet investigation for Social Webnet examination for criminal looks to address. As the creator puts it, was brought into the world in a drive; I was profoundly disturbed about Facebook sharing/trade my data and with the overall absence of privacy on the unmistakable internet. Social Webnet investigation, before Facebook, surrendered it to clients to choose whether and this present reality character of Facebook has made the way of life incredibly appealing to every one of those in the merchandise who search precisely in their focus on publicizing [37].

A. Bitcoins Scam

Touch coin is digital money utilized financially on the DW. It is likewise sensible money for cybercriminals. Onion Cloning and Proxying are the instances of such misconducts as Figure 9.5 below shows the use of money transfer for illegal activities across the boundaries using virtual currency.

Figure 9.5 Bitcoins scam on the DW.

B. Child Pornography

The investigation found that youngster erotic entertainment produces the most traffic to the shrouded locales on concealed web path. It is difficult for a normal client to discover such destinations. It is a demonstration that misuses the youngsters for sexual incitement and maltreatment of kids during sexual acts. It additionally incorporates the sexual pictures of kid's erotic entertainment. The Lolita City [32] is used for illicit and morphed photographs and harvesting of youngster sexual entertainment [39] and has around 15,000 individuals.

C. Hired Assassin

Likewise, DW is a stage for assassin recruitment. That is where it is possible to hire an expert assassin. When the named programmer infiltrated the Besa-Mafia website and posted the material online. The material published includes client pages, individual messages, 38 hit orders, and an envelope composed of approximately 200 pictures of the casualty, some of the homicide for enlisting bunches to include Unfavorable Options, Web-net hitmen, and C'thulhu [35]. Hostile Solution accepts the Bit-coin installment as it was. Web-net employed gunmen who profess to be a triplet of executioners working in the Foreign Exchange Merchandise deal give a commission to appeal to their friends [31]. DW hired gunmen are contract executioners and professional killers on the profound web who takes orders from the individuals and acknowledges installments in cryptographic forms of money including Bit-coin. The individual who will do the request is known as an assassin; the service given to you using the web forum is known as the contract killers services.

D. Drug Harvesting

DW is an illicit dispensary of prohibited and hazardous substances [40] sold in exchange for cryptographic forms of currency: Bitcoin, Ethereum [14] and wave, and so forth, for example. Also, Silk Road [41] was one of the renowned commercial centers for counterfeit narcotics and unlicensed drugs. The FBI shut down this website in 2013 [6]. The Merchandise Location is a website that was closed down a year ago. Alpha Bay is the main commercial hub for drugs.

E. Frauds

Checking cheats alludes to the taking and trade [37] of client's MasterCard qualifications and individual data. It is the most widely recognized sort of misconduct that occurs on the DW. Several reasons make this extortion generally well known over the DW. Dark Net merchandise offers credit and checks cards. The destinations have various URLs that are immediate for the client to a similar page. The pedlar from different sorts of gatherings posts advertisements in which they indicate, the various discussions additionally give a live office visit. Merchants offer cards at lower costs. A couple of money moves stages are additionally vulnerable to checking cheats.

F. Human Harvesting for Illicit Trading

Dark Market is a spot on the DW where human harvesting happens. The majority of the overcomers of human harvesting were selected for sex traffics and work harvesting. Figure 9.6 below shows the Human harvesting merchandise structure and connections on the DW platform.

Different reports have indicated that DW has assisted with pushing this misconduct more profound into the mystery.

G. Information Leakage

For example, the helpful tools for whistleblowers, advocates, and law enforcement are various levels of hidden network paths that underpin anonymity. DW is also a stage for the publication of classified information by programmers. If the Visa accounts have been posted, a bunch of programmers log in to about 32 million user accounts as a 9.7 GB data dump on the DW. Similarly, over 1.4 billion individual harvests [42] were distributed over the DW in 2017 as plain material that was readily accessible on the web. Indeed, even the DW centers pay the staff to release the data from the organization.

H. Hidden Web Trade Replica

Onions cloning is like a proxy solution. The scam artist makes the genuine site or website replicate and changes the links so that the user is

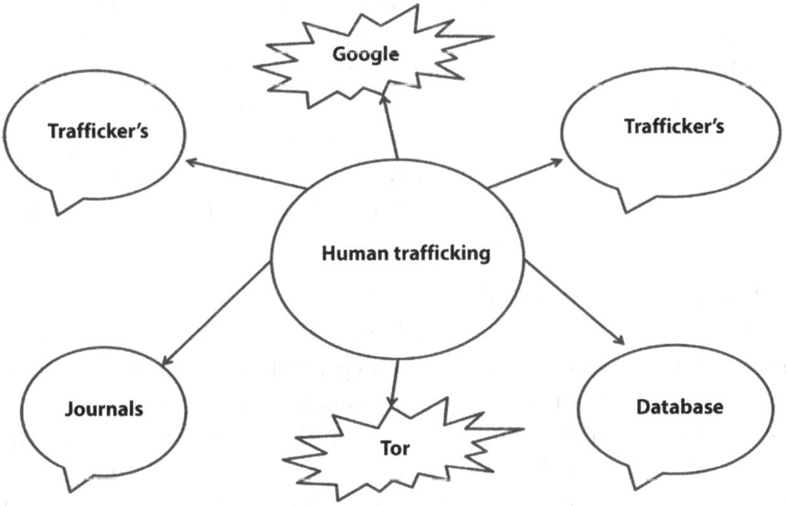

Figure 9.6 Human harvesting is sold on the DW.

coordinated to their fooled positions to steal the money from the customer side. Europol officials shared their concern that Bitcoins had started to play a developing role in illegal operations. Since its formation in 2014, DDoS-Bit-coin, a community of cybercriminals behind Distributed Denial of Service (DDoS) attacks, has targeted more than 140 syndicates. This also triggers numerous meetings and prompts cyber extortion. The DDoS-Bit-coin community previously undermined casualties via email with a DDoS attack until emancipated as Bit-coins are not paid, according to the Europol authorities. In comparison, the emergence of Bit-coin encourages the rise of Cyber-hijackers in the DW world.

I. Revenge Porn

Pornography of Revenge is the distribution of sexual imagery and the harvesting of individuals without their consent. Several destinations were closing down. Google has just removed the revenge pornography destinations from its query products when it is listed, on the off chance that speak of the surface network. Before switching to the DW, Pink-Meth was originally the website of the Surface Network. It helps consumers to anonymously present their substances.

J. Maltreatment

Locations in the Red Room are those destinations where viewers spend millions to watch streaming homicides, attacks, adult entertainment for children, and different kinds of torture. Even now, there is also no indication that they exist. In the off case that they do exist, they cannot be opened at that stage by hidden network paths for web-based live harvesting is unnecessarily delayed. According to a few reports, customers of a pedophile site charged a colossal sum of money to witness Scully's maltreatment and maltreatment of a little youngster being harvested for one, it was an agreement given by Scully's company, No Constraints Nice. Daisy's Ruin, one of his harvests, was extensively spoken about at the event, which was not a sniffed video, but rather contains actual rape and disturbing exploitation of a young child [34–36].

K. Cyber-Attacks by Social Web Net

Cyber-attacks that can be dispatched via DW. The DW's biggest weakness is its anonymity, which increases the degree of certainty of the network danger and can rely on losses without much of a stretch attack [7]. The attacker manages the first and last router in the hidden web path network in this kind of attack, using the circumstances and data highlights to relate the streams over certain routers to destroy the confidentiality of the TOR. Previously, various administration agencies had the option to pulverize the privacy of numerous clients with the assistance of partnership abuse. Since it is an extremely refined numerical method, there is no elective tactic to avoid this form of attack. This kind of attack is not only for programming but is used against clients at the same time [30].

9.7 DW—Frontier of Illicit Human Harvesting

The part of the internet that can be easily accessed by web crawlers is where the vast majority of our everyday internet movement takes place. There is something completely more about the web than meets the eye, be that as it may. For the Internet, the size of the surface network must be greater than many times [41]. This is known as the deep web which includes a piece of the web that is not visible. Another subsection of the site called the DW sits below this. It is the portion of the internet that is intentionally covered up and offers the perfect protection for human harvesting. Using multi-layered codification technologies called onion codification, the DW is close to maintaining privacy for its customers. Admission to the onion-encoded web forum (with the .onion domain) is transmitted across layers like an onion, as the name implies. The defining address of a device is scrambled and transferred to another voluntary worker at each layer of the IP address of the customer, and henceforth the following layer. The website never even realizes who they are referring to since the goal website just gets the last IP address in the course cycle.

An easily available web index named Concealed web path [42] may be used to get to the DW. Without whom they are, Tor investigates these onion layers and leads the searcher to the perfect website. Concealed web route does not continue to harvest searches, because all things considered at this point are not a hint of motion.

9.8 Organ Harvesting Apprehension

To obtain an understanding of the uncertain cycle and impact of several credits on the harvesting of human organs, this section reviews current research [45] and knowledge obtained between meetings with subject specialists. Because of its temperament, the harvesting of human organs requires various teamwork. In any event, in the computerized environment, a large part of these chapters handles the means executed and the follows left. By the way, zeroing on these follow-ups and observing the resulting electronic communications chains will include signs that will help and prompt certifiable tests. Several persons with different foundations are concerned with human organ follow-up. The organ tracking mechanism and their interactions are seen in Figure 9.7.

The accompanying passages besides portray them just like their inspirations:

- **Buyers**: Human organ patients are not patients with confusion from more lavish nations souring from illnesses, such as renal deficiency, for example. Despite the restricted limits of storage and transportation, organs are acquired to sell them after withdrawal, take all into account. Or maybe, on the other hand, customers and dealer patrons are taken by agents to the comparative region, and the move is carried out by professionals who often profit from the theory in fiscal terms.
- **Pedlar/Donors:** Organ pedlars are typically poor, young people. Although a few vendors are proactively looking for activities to sell their organs cash to receive (either individually or persuaded by their monetary atmosphere), some are drawn in by oaring cash plugs without explicitly signaling that an organ departure is a true return. Energetic outcasts who cover their expenses with organs are another continuously rising pack.
- **Brokers:** Brokers are the central broker who ties the entire organ together following the web-net. Also, they benefit most financially from the get-together. Although merchants get a four-to-a-low five-digit whole in US dollars, buyers usually pay for an organ a six-digit whole in US dollars.

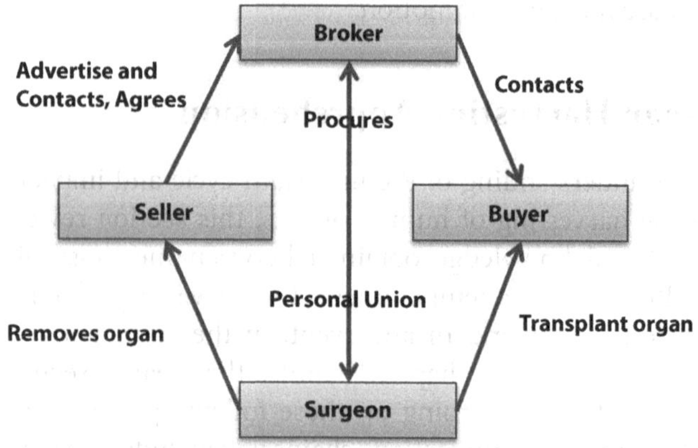

Figure 9.7 The organ tracking process and its relationships.

- **Transplant Surgeons/Hospitals:** While breaching authentic and moral norms, some clinical experts transfer pursued organs excitedly, benefiting from the over-the-top costs of organs.

The process depicted so far relies on social connections between brokers, surgeons, and their current as well as previous patients that are largely resolved. Correspondence between such meetings is generally confidential and cannot be publicly examined. There are also praiseworthy cases in which performers need to impart their cravings to a wider audience, in any event. The Organ Monitoring Mechanism Summary is given in Figure 9.8 below.

- A sick patient urgently looking for an explicit organ and has not set up any associations with any individual who is as of now associated with organ follows.
- A broker that already has a prospective buyer does not have a pedlar available yet. Since organ follow-up cannot be conveniently marketed on the surface web, more private communication networks are used by all meetings of

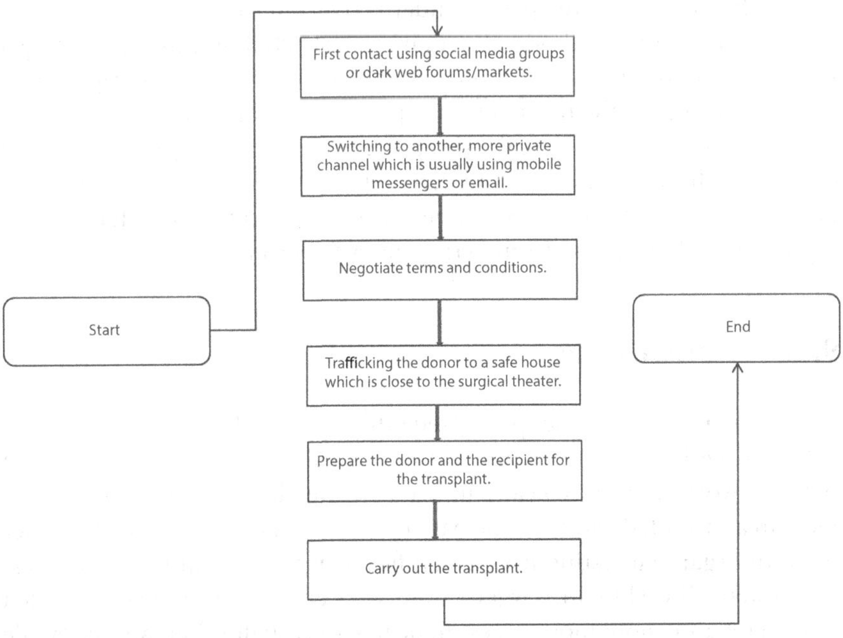

Figure 9.8 Overview of Organ Tracking Process.

actors, putting social media (profound web) and DW into operation. For this case, the loop can be outlined as follows (approximately):

9.9 Result and Discussions

Having taken a gander at the writing about concealed web path concealed services, the accompanying research holes have been recognized and proposed. Cybercriminals have profited by shrouded services via having the option to purchase misuses—including zero-day—from commercial centers, draw in with different criminals on gatherings, have order and control workers for botnets, and exfiltration malware, and recruit proficient criminals or their abilities as-a-service [36]. The anonymity trait the DW presents to its clients makes these exercises conceivable and empowers cybercriminals to lead their issues.

There are different kinds of web crimes and actors present, from the beginner 'content youngsters' utilizing promptly accessible instruments and adventures to target profoundly weak systems to the exceptionally engaged and gifted Advanced Persistent Web peril known to be behind huge instances of undercover work against huge syndicate and country states. The Persistent Web peril wonder is referred to utilize zero-day misuses as one of the methods of acquiring restricted admittance to target systems [45]. The inventive strategies of Persistent Web peril represent a critical test to even the most develop episode reaction groups [44]. Be that as it may, there was no critical writing found unequivocally investigating a connection between the DW and Persistent Web peril marvels, and whenever researched it can conceivably be abused in checking or better protecting against cyber-attacks from Persistent Web peril later.

9.10 Conclusions

This chapter examined the specialized difficulties, and express that the DW is loaded with an unlawful and hazardous substance. DW is the part of the web world that is typically utilized by the clients to do some movement in a shrouded way without leaving any follows. It has become a center point of organ harvesting uses a significant part of the innovation accessible today. The Human (Organ) traffickers are effectively using Hidden web secret and anonymous access to deal with casualties' situations. While endeavors battle human (Organ) harvesting as a severe cyber threat and crime, the investigation requires further comprehension and tracking of malicious cyber activities and chaining of cyber users involved in illegal trade.

References

1. Rajawat, A. S., Rawat, R., Mahor, V., Shaw, R. N., Ghosh, A., Suspicious big text data analysis for prediction—on darkweb user activity using computational intelligence model, in: *Innovations in Electrical and Electronic Engineering*, pp. 735–751, Springer, Singapore, 2021.
2. Weimann, G., Going dark: Terrorism on the dark web. *Stud. Confl. Terror.*, *39*, 3, 195–206, 2016.
3. Gokhale, C. and Olugbara, O.O., Dark Web Traffic Analysis of Cybersecurity Threats Through South African Internet Protocol Address Space. *SN Comput. Sci.*, *1*, 5, 1–20, 2020.
4. Gerbaka, B., Richa, S., Tomb, R., Commercial sexual exploitation of children and trafficking, in: *Child Sexual Abuse, Exploitation and Trafficking in the Arab Region*, pp. 101–154, Springer, Cham, 2021.
5. Heinl, M.P., Yu, B., Wijesekera, D., A Framework to Reveal Clandestine Organ Trafficking in the Dark Web and Beyond. *J. Digital Forensics Secur. Law*, *14*, 1, 2, 2019.
6. Ozhathil, D. K., Tay, M. W., Wolf, S. E., Branski, L. K., A narrative review of the history of skin grafting in burn care. *Medicina*, 57, 4, 380, 2021.
7. Al-Dayel, N., Anfinson, A., Anfinson, G., Captivity, migration, and power in Libya. *J. Hum. Trafficking*, 1–19, 2021.
8. Rawat, R., Rajawat, A. S., Mahor, V., Shaw, R. N., Ghosh, A., Dark web—onion hidden service discovery and crawling for profiling morphing, unstructured crime and vulnerabilities prediction, in: *Innovations in Electrical and Electronic Engineering*, pp. 717–734, Springer, Singapore, 2021.
9. Rawat, R., Mahor, V., Chirgaiya, S., Shaw, R. N., Ghosh, A. Analysis of darknet traffic for criminal activities detection using TF-IDF and light gradient boosted machine learning algorithm, in: *Innovations in Electrical and Electronic Engineering*, pp. 671–681, Springer, Singapore, 2021.
10. Berghel, H., Which Is More Dangerous—The Dark Web or the Deep State? *IEEE Ann. Hist. Comput.*, *50*, 07, 86–91, 20172017.
11. Kar, D. and Spanjers, J., *Transnational Crime and the developing world*, Global Financial Integrity. Retrieved on May, 30, Washington, 2019, 2017.
12. Rawat, R., Mahor, V., Rawat, A., Garg, B., Telang, S., Digital transformation of cyber crime for chip-enabled hacking, in: *Handbook of Research on Advancing Cybersecurity for Digital Transformation*, pp. 227–243, IGI Global, 2021.

13. Mann, S., Naylor, K.L., McArthur, E., Kim, S.J., Knoll, G., Zaltzman, J., Treleaven, D., Ouedraogo, A., Jevnikar, A., Garg, A.X., Projecting the Number of Posttransplant Clinic Visits With a Rise in the Number of Kidney Transplants: A Case Study From Ontario, Canada. *Can. J. Kidney Health Dis.*, *7*, 2054358119898552, 2020.
14. Kovacikova, L., Bena, L., Lackova, E., Zilinska, Z., Granak, K., Skalova, P., Vnucak, M., Laca, L., Mokan, M., Dedinska, I., Gender Differences in Kidney Transplantation—10-Year Retrospective Study. *Acta Med. Martin.*, *18*, 3, 24–29, 2018.
15. Rawat, R., Mahor, V., Chirgaiya, S., Rathore, A. S., Applications of social network analysis to managing the investigation of suspicious activities in social media platforms, in: *Advances in Cybersecurity Management*, pp. 315–335, Springer, Cham, 2021.
16. Serebrennikova, A.V., Minyaseva, T.F., Kala, N.S., Malinovsky, A.A., Malinovskaya, V.M., Grynchak, S.V., Comparative analysis of foundations of legal regulation of criminal liability for organ trafficking in the Russian Federation, Kazakhstan, and the European Union. *J. Adv. Res. Law Econ.*, *11*, 1405, 2020.
17. Rawat, R., Rajawat, A. S., Mahor, V., Shaw, R. N., Ghosh, A., Surveillance robot in cyber intelligence for vulnerability detection, in: *Machine Learning for Robotics Applications*, pp. 107–123, Springer, Singapore, 2021.
18. Kar, D. and Spanjers, J., *Transnational crime and the developing world*, Global Financial Integrity. Retrieved on May, 30, Washington, 2019, 2017.
19. Rajawat, A. S., Rawat, R., Barhanpurkar, K., Shaw, R. N., Ghosh, A., Vulnerability analysis at industrial internet of things platform on dark web network using computational intelligence. *Comput. Intell. Syst. Appl.*, 39–51, 2021.
20. Raza, H. and Raza, M. R., A study of blockchain technology, bitcoin and other cryptocurrencies as means of money laundering, frauds and scams. *Global Med. Soc. Sci. Res. J. (GMSSRJ)*, 2, 1, 73–84, 2021.
21. Adelina, A. and Ishadi, N . K., The act of abduction and women trafficking in non-international armed conflict (NIAC): Boko haram case. *Padjadjaran J. Int. Law*, 5, 1, 81–97, 2021.
22. Loesing, K., Sandmann, W., Wilms, C., Wirtz, G., Performance measurements and statistics of Tor hidden services, in: *2008 International Symposium on Applications and the Internet*, pp. 1–7, IEEE, 2008, July.
23. Adler, M., An analysis of the degradation of anonymous protocols, in: *Network and Distributed Security Symposium*, 2012.
24. Bauer, K., McCoy, D., Grunwald, D., Kohno, T., Sicker, D., Low-resource routing attacks against Tor, in: *Proceedings of the 2007 ACM Workshop on Privacy in Electronic Society*, pp. 11–20, 2007, October.
25. Yao, G., Bi, J., Vasilakos, A.V., Passive IP traceback: Disclosing the locations of IP spoofers from path backscatter. *IEEE Trans. Inf. Forensics Secur.*, *10*, 3, 471–484, 2014.

26. Davis, S. and Arrigo, B., The Dark Web and anonymizing technologies: legal pitfalls, ethical prospects, and policy directions from radical criminology. *Crime, Law and Social Change*, 1–20, 2021.
27. Christin, N., Traveling the Silk Road: A measurement analysis of a large anonymous online marketplace, in: *Proceedings of the 22nd International Conference on World Wide Web*, 2013, May, pp. 213–224.
28. Dolliver, D.S., Evaluating drug trafficking on the Tor Network: Silk Road 2, the sequel. *Int. J. Drug Policy*, *26*, 11, 1113–1123, 2015.
29. Kumar, A., Abhishek, K., Nerurkar, P., Khosravi, M. R., Ghalib, M. R., Shankar, A., Big data analytics to identify illegal activities on Bitcoin Blockchain for IoMT. *Pers. Ubiquit. Comput.*, 1–12, 2021.
30. Pay, M.T., Gangoiti, G., Guevara, M., Napelenok, S., Querol, X., Jorba, O., Pérez García-Pando, C., Ozone source apportionment during peak summer events over southwestern Europe. *Atmos. Chem. Phys.*, *19*, 8, 5467–5494, 2019.
31. Gallagher, K., Patil, S., Dolan-Gavitt, B., McCoy, D., Memon, N., Peeling the Onion's User Experience Layer: Examining Naturalistic Use of the Tor Browser, in: *Proceedings of the 2018 ACM SIGSAC Conference on Computer and Communications Security*, 2018, October, pp. 1290–1305.
32. Westlake, B., Bouchard, M., Frank, R., Assessing the validity of automated webcrawlers as data collection tools to investigate online child sexual exploitation. *Sex Abuse*, *29*, 7, 685–708, 2017.
33. Rajawat, A. S., Rawat, R., Shaw, R. N., Ghosh, A., Cyber physical system fraud analysis by mobile robot, in: *Machine Learning for Robotics Applications*, pp. 47–61, Springer, Singapore, 2021.
34. Farwell, J.P., Strengthening Private–Public Partnerships in National Cybersecurity, in: *Understanding Cybersecurity: Emerging Governance and Strategy*, p. 205, 2018.
35. Fossum, L., Cyber Conflict Bibliography, 2015 Update, in: *GWU Law School Public Law Research Paper*, (2015 57), 2015.
36. Jardine, E., The Dark Web dilemma: Tor, anonymity and online policing, in: *Global Commission on Internet Governance Paper Series, (21)*, 2015.
37. Westlake, B., Bouchard, M., Frank, R., Comparing methods for detecting child exploitation content online, in: *2012 European Intelligence and Security Informatics Conference*, 2012, August, IEEE, pp. 156–163.
38. Kara, S., Supply and demand: Human trafficking in the global economy. *Harvard Int. Rev.*, *33*, 2, 66, 2011.
39. Nagle, L.E., Selling souls: The effect of globalization on human trafficking and forced servitude. *Wis. Int. Law J.*, *26*, 131, 2008.
40. Rajawat, A. S., Rawat, R., Barhanpurkar, K., Shaw, R. N., Ghosh, A., Blockchain-based model for expanding IoT device data security. *Adv. Appl. Data-Driven Comput.*, 61, 2021.
41. Rawat, R., Mahor, V., Chirgaiya, S., Shaw, R. N., Ghosh, A., Sentiment analysis at online social network for cyber-malicious post reviews using machine learning techniques. *Comput. Intell. Syst. Appl.*, 113–130, 2021.

42. Adler, M., An analysis of the degradation of anonymous protocols, in: *Network and Distributed Security Symposium*, 2012.
43. Mirea, M., Wang, V., Jung, J., The not so dark side of the darknet: a qualitative study. *Secur. J.*, *32*, 2, 102–118, 2019.
44. Bele, J. L., Cryptocurrencies as facilitators of cybercrime, in: *SHS Web of Conferences*, vol. 111, p. 01005, EDP Sciences, 2021.
45. Miller-Perrin, C. and Wurtele, S.K., Sex trafficking and the commercial sexual exploitation of children. *Women Ther.*, *40*, 1–2, 123–151, 2017.
46. Mageshkumar, C., Rajeshkumar, J., Atlas, L.G., Hamsagayathri, P., Blockchain in 5G, in: *Cryptocurrencies and Blockchain Technology Applications*, pp. 43–54, 2020.
47. Mazi, H., Arsene, F. N., Dissanayaka, A. M., The influence of black market activities through dark web on the economy: a survey, in: *The Midwest Instruction and Computing Symposium (MICS)*, Milwaukee School of Engineering and Northwestern Mutual, Milwaukee, Wisconsin, April, 2020.
48. Day, T., Dennis, K., Gibson, H., TENSOR: A solution to dark web investigations, in: *Dark Web Investigation*, pp. 119–135, Springer, Cham, 2021.
49. Holland, K. Trading Life: Organ Trafficking, Illicit Networks, and Exploitation. NYUJ Int'l L. & Pol., 53, 717, 2020.

10

Deep Learning Techniques for Data Analysis Prediction in the Prevention of Heart Attacks

C.V. Aravinda[1*], Meng Lin[2†], Udaya Kumar[3], Reddy K.R.[3] and G. Amar Prabhu[4]

[1]*N.M.A.M. Institute of Technology Nitte, Karkala, India*
[2]*Dept. of Electronic and Computer Engineering (The Graduate School of Science and Engineering), Ritsumeikan University, 1-1-1 Noji-higashi, Kusatsu Shiga, Japan*
[3]*Department of Information Science Engineering, Dayananda Sagar College of Engineering, Shivage Malleshwara Hills, Bengaluru, Karnataka, India*
[4]*Komatsu Kaihatsu Company, Otemachi Kariya-shi, Achi-Ken, Japan*

Abstract

Coronary illness may have an awful scope of divergent conditions that could influence your heart. It is one of the furthermost mind-boggling disorders to foresee a given number of components in your body that can prompt it. Distinguishing and anticipating it represents a lot of challenge for both specialists and analysts. If we extricate the bereavement caused between the avoidable death and the other death in India, the unnecessary death stands in the 3rd position due to Heart-attack. Across the globe, these types of deaths can increase to more than 23.6 million by 2030. Around 80% of deaths by heart attack mainly affects the younger generation. This research article not only supports doctor specialists in detecting and foreseeing heart disease by attaining precision levels but also advises the effective medicine for the disease. Given sensor data, the deep learning algorithms are applied along with Neural Network, Random Forest and Decision Tree classifier for analyzing the patient's data to predict heart disease. The experiment shows that prediction of heart disease gives promising results of about 90% accuracy.

**Corresponding author*: aravinda.cv@nitte.edu.in
†*Corresponding author*: menglin@fc.ritsumei.ac.jp

Kavita Sharma, Yogita Gigras, Vishnu Sharma, D. Jude Hemanth and Ramesh Chandra Poonia (eds.) Internet of Healthcare Things: Machine Learning for Security and Privacy, (217–240) © 2022 Scrivener Publishing LLC

Keywords: Neural-Network, artificial neural network, Random Forest, decision tree, Naïve Bayes

Abbreviations

LMIC Low and Middle-income countries;
KNN K-Nearest Neighbor;
RF Random Forest;
NN Neural Network;
CBR Case Based Reasoning;
RFA Random Forest Algorithm;
DTA Decision Tree Algorithm;
ANN Artificial Neural Network.

10.1 Introduction

Cardiovascular disorder has become the predominant undertaking for the healthcare unit (doctors, medical centers, hospitals). Cardiovascular disease is a sickness elegance which includes heart and blood vessels. The American Coronary heart syndicate has appraised that 17.3 million human beings are losing their existence because of cardiovascular disease in line with 12 months, especially coronary heart assaults, strokes, coronary heart sickness, pulmonary coronary heart ailment, etc. This planetary cause of death can upturn the total to grow more than 23.6 million by 2030. Due to this heart failure, the major countries with high population and poor income communities are affected more and the death rate is around 80% especially in the younger age generation. To control this cause, there is much need for health support such as diagnosis and prediction, but this cannot be considered as a complete solution for accuracy in the verdict. Plenty of exploration is being done for the diagnosis of heart disease but still, the impediments in various factors are causing deferral in the diagnosis of heart disease and deciding the accuracy [1]. In India, the generation of medical image data is estimated at 100 GB every year. Computer-based patient records also generate different clinical information such as information of patient's, diagraphs, diagnosis, procedure, medicines, and laboratory information. Patients also generate data such as arm guard generating data and other wearable components. In addition, there are real time data sources from blood pressure devices, temperature, and the Intensive Care Unit. This research focuses on considering the wide variety of data and hence the scalable Machine Learning (ML) and Data Mining (DM) algorithms have been considered for processing data. The enormity and complexity of these datasets present great challenges in analytics and subsequent applications to a practical clinical environment. This research is all about finding the percentage of the threat of having affected by uneven breathing because of excessive excess cholesterol in addition to other fundamental physical signs and symptoms [2]. As per WHO, ischemic and hypertension attack is one among the prime causes of death. This infection will become existence-threatening if not addressed correctly. Therefore, the proposed method attempts to address the issues as detailed below.

- It has many potential ideologies, diverse clinical features and numerous clinical subsets.
- If the model can detect heart failure earlier, there may be a potential chance to reduce the cost of hospitalization associated with heart failure.
- It introduces the slowdown of progression of heart failure as early intervention, improves life in terms of quality and mortality reduction.
- Improves heart failure prevention as compared to the existing clinical guidelines.

The rest of the article is organized as follows. Section 10.2 briefs the related prediction work of various authors studied and listed in the literature survey. Sections 10.3 describes the proposed method. Section 10.4 describes various algorithms for data pre-processing, feature extraction, and data cleaning. Section 10.5 describes the data preparation. Section 10.6 delivers the results of experiments and comparison details. Section 10.7 provides a conclusion and summary of the work carried out so far.

10.2 Literature Survey

In this section mainly we are focusing on the sensor datasets mostly which are based on heart failure disease data which are freely available in cloud repositories. Since we have incorporated a similar problem and achieved good results and came out with satisfactory results in our paper, many of the researchers have considered these data for the prediction of heart disease for prevention purposes. The repository statistics sets consist of speculation factors (functions) inclusive of "age", "sex", "chest ache", "blood strain", "cholesterol", "blood sugar", "electrocardiograph consequences", "maximum heartbeat rate", etc [3].

Vijayashree *et al.* [4] identified only a few parameters for data analyzing and concluded to form a proficient quick reliable mechanized system that may envisage the heart disease early stage.

Zhu *et al.* [5] implemented a diverse system of studying strategies and related the overall concert on eight clinical datasets the usage of five specific parameters like "enactment", "pellucidity", "clarification", "reduction", and "missing data handling".

Soni *et al.* [6] stated that the awareness is on the use of special processes and groupings of numerous goal attributes for smart and powerful coronary heart assault prediction the use of statistics mining like ANN, Soft computing rules, etc. The results obtained from the statistics mining technique is equal to the fact of the decision tree, meanwhile, Bayesian is also having a similar accuracy as the decision tree result. The predictive techniques like K-Nearest-Neighborhood, Neural Networks are based on clustering.

Austin *et al.* [7] specified that doctors categorize patients based on the disease irrespectively. In addition to this regular enthusiasm for gathering patients with different etiologies has been done. The decision tree algorithm techniques are used to classify patients frequently. Authors claimed the results with that of conventional classification techniques to group patients with the cardiac breakdown as indicated by the add-on sub-types like fortified part. Finally, they concluded conventional regression is better for expecting the probability of Heart Failure concerning other methods proposed earlier.

Chowdhury [8] denotes the use of "Artificial-Neural-Network" to predict the unhealthiest credentials at an early stage. As mentioned in [8], it involves the preparation of a multilayer perceptron with a Back-Propagation rule to recognize a pattern for the prophecy of infant diseases.

Kadi *et al.* [9] added that the enthusiasm behind this paper is to survey papers relating to the use of Data Mining procedures in cardiology, to sum up, and examine proof in regards to the Data Mining strategies most as often as possible utilized in cardiology; the presentation and examinations of Data Mining models in cardiology. Neural systems, decision trees and support vector machines (SVM) were recognized when creating Data Mining models in cardiology.

Han *et al.* [10] collected the stored data from various domains which had the factors of cardiovascular disorders and developed a framework. This framework consisted of nutrition data from the Korea center for disease control and prevention center. These data were mapped with a big data concept using Hadoop and map-reduce nodes for parallel cluster nodes. The fact that evaluation became composed to evaluation layer for the expectation model was primarily based on the improved evaluation pipeline.

10.3 Materials and Method

The idea of the proposed method is to develop a predictive model for heart failure that is helpful in the early detection in humans. This is shown in Figure 10.1.

The focus on predicting the onset of heart failure is an interesting and potentially possible target. In the prospective study, the first step is to identify the cohort of patients, then deciding on what and how information can be collected, then start the data collection. In contrast, in a retrospective study, firstly to identify the patient's cohorts from existing data, likewise to consider the past sensor records of patients, then retrieve all the data about the cohort. Consequently, in a retrospective study, the data set already exists and hence just need to identify the right subset and retrieve them. In a cohort study, the goal is to select a group of patients who are exposed to a particular risk. For example, to predict the model for predicting heart failure readmission (i.e., heart failure patient after discharged from the hospital comes back again to the hospital due to heart failure) [11–15].

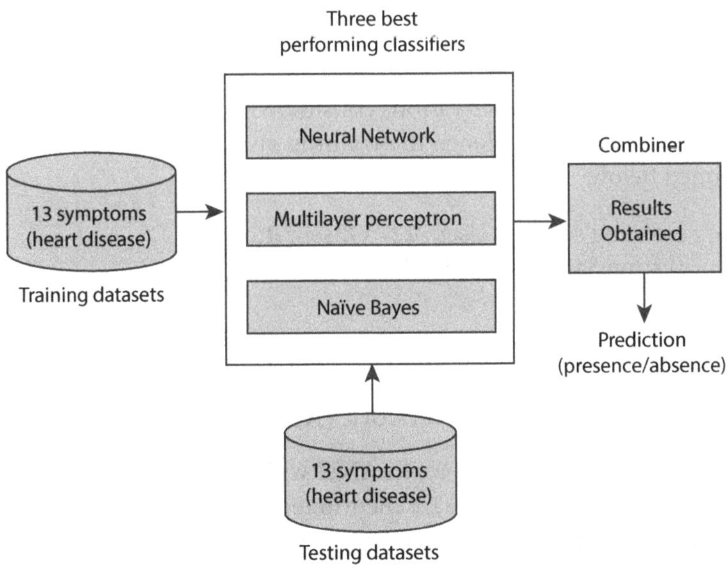

Figure 10.1 Architecture of the proposed method.

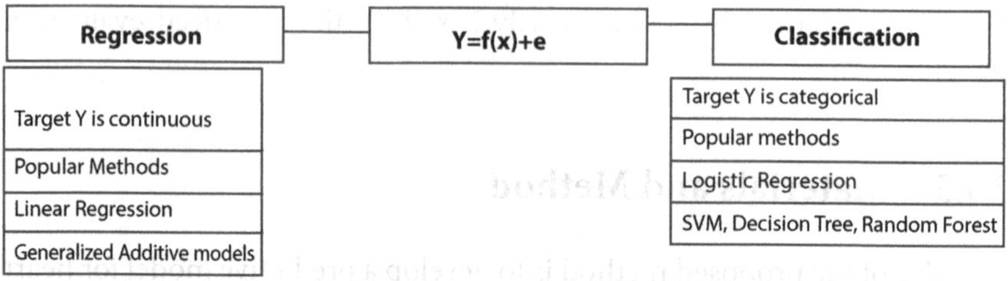

Figure 10.2 Predictive models.

The key cohort study is to define the right inclusion and exclusion criteria to figure out what patient to include. Prospective study versus retrospective and Cohort study versus case-control. Depending upon the combinations, there are four different options: Prospective-cohort, Prospective-case-control, Retrospective-cohort, and Retrospective-case-control as shown in Figure 10.2.

10.3.1 Cohort Study

The group of patients with the risk of being exposed.

1. Target: Heart failure Readmission.
2. Cohort: All the patients discharged from the hospital
3. Key: Define the right inclusion/exclusion criteria.

The first 13 rows of Table 10.1 are dependent variables and row 14 is an independent variable [16–20].

The total number of observations considered are 303 in the dataset, 25% was split for validation, Hence the training and testing samples used are mentioned below:

- Training samples Used: 227 Observations
- Testing Samples Used: 176 Observations

10.4 Training Models

10.4.1 Artificial Neural Network (ANN)

The training and testing samples in ANN with 500 Epochs are considered for training. The ANN details are mentioned in Table 10.2. The activation functions considered are rectified linear activation function (ReLu) and Sigmoid [21–25].

Table 10.1 Important factors extracted for training.

Column number	Data explanation
"One–Three"	A Period of Human Life
"Two–Four"	1 = "Male"; 0 = "female"
"Three–Nine"	Value 1: "myocardial-infraction" Value 2: "non-myocardial-infraction" Value 3: "abnormality"
"Four–Ten" (test-bps)	"resting-Blood-Pressure"
"Five–Twelve" (cholesterol)	"serum-sterol"
"Six–Sixteen" (fbs)	"fasting-Blood-Pressure >120 mg/dl)" 1 = true; 0 = false
"Seven–Nineteen" (restecg)	"Value 0: normal" "Value 1: having-myocardial-ischemia (T wave inversions >0.05 mV)" "Value 2: showing probable or definite left ventricular hypertrophy by Estes"
"Eight–Thirty-Two" (thalach)	"maximum-heart-rate-achieved"
"Nine–Thirty-Eight" (exang)	"exercise-induced-angina" (1 = yes; 0 = no)
"Ten–Forty" (old peak)	"ST depression induced by exercise relative to rest"
"Eleven–Forty-One" (slope)	Value 1: "upsloping" Value 2: "flat" Value 3: "down sloping"
"Twelve–Forty-Four" (ca)	"number-of-major-vessels (0–3) colored by fluoroscopy"
"Thirteen–Fifty-One" (thal)	3 = "normal"; 6 = "fixed defect"; 7 = "reversable defect"
"Fourteen–Fifty-Eight" (num)	"predicted-attribute" Value 0: <"50\% diameter narrowing" Value-1: >"50\% diameter narrowing"

Table 10.2 Layer details.

Layer name	Number of neurons with activation function
"Input Layer"	8 (ReLu)
"Hidden Layer"	16 (ReLu)
"Hidden Layer"	32 (ReLu)
"Output Layer"	1 (Sigmoid)

10.4.2 K-Nearest Neighbor Classifier

The K-Nearest Neighbor (KNN) is a supervised machine learning algorithm. The set of rules of KNN assumes that similar things are near to each other. KNN also assumes that the data points that are close to each other must be of the same class. To calculate the distance between the data points, various distance formulas can be used including Euclidean, Manhattan, etc [26–30]. The other parameter required for KNN is the number of points to consider the new data point as belonging to the trained classes. The same dataset was passed through a K Nearest Neighbors Classifier. The KNN details are mentioned below:

- N Neighbors: 3
- Distance Used: Euclidean

In order not to face a dominance problem in finding distance, feature scaling is done to reduce all parameters to a range of −2 to +2. as shown in Figure 10.3. From the matrix, the validation accuracy obtained is 82.89%.

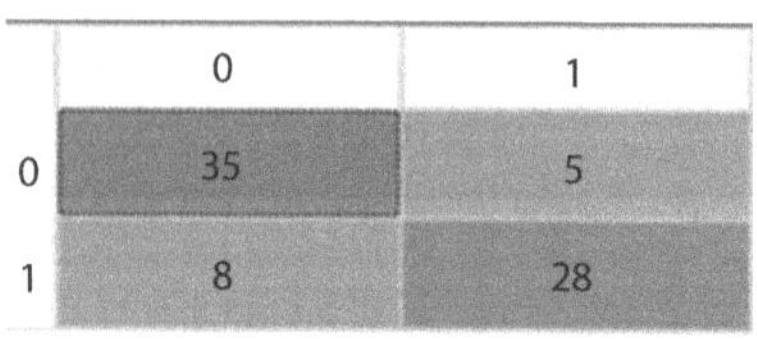

Figure 10.3 Obtained confusion matrix for K Nearest Neighbors Classifier.

10.4.3 Naïve Bayes Classifier

This Naïve Bayes model is easy and useful for huge data sets and this known to perform better classification methods. The equation is as follows "P(c / x) = P(x / c) P(c) / P(x)".

The experiment was conducted based on this algorithm for the classification and prediction of heart attack based on various symptoms like diabetes and high blood pressure.

P (Yes | diabetic) = P (diabetic | Yes) ∗ P(Yes) / P (diabetic). This can be read as the probability of event diabetic given that event blood pressure has occurred is equal to the probability of blood pressure given that diabetic has occurred times and the probability of event diabetic by probability of event blood pressure [31–35]. This method was used to predict the probability of diverse classes on numerous features.

$$P(Y = Yes/X = diabetic, blood - pressure)^\wedge = P(Y = Yes) * P(X = ((diabetic, blood - pressure))/Y = Yes)P(X = (diabetic, blood - pressure))^\wedge$$

For the experiment purpose, the equation\ref{equ:1} is expanded to contain only basic probabilities

$$P(Y = Yes/X = diabetic, blood - pressure)^\wedge = P(Y = Yes) * P(X = ((diabetic,))/Y= Yes)P(X = (blood\ pressur/Y = Yes)P(X = (diabetic, blood - pressure))^\wedge$$

based on Equation (10.4.3), the probability is calculated.

$$P(Y = Yes) = count(Yes) + count(No)$$

The confusion matrix obtained is shown in Figure 10.4.

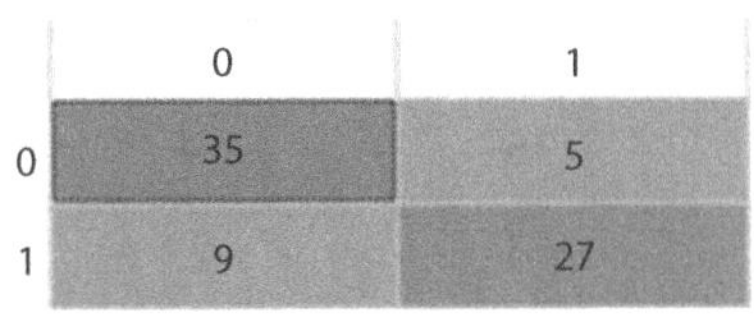

Figure 10.4 Obtained confusion matrix for Naïve Bayes Classifier was 81.57%.

10.4.4 Decision Tree Classifier (DTC)

The decision tree algorithm belongs to supervised learning algorithms. This algorithm is used for fixing regression type problems. This algorithm model makes the predictions by self-importance a series of trials on a certain argument. The main intention to use this DTC is to predict the patient's data to identify the reason for suffering mainly from these parameters. In these cases, doctors will not have the idea to conclude for what reason this attack took place [36–40]. Therefore, build a decision tree to predict the patient having attack symptoms. Hence, in this case, the values for the continuous variables are predicted and the obtained accuracy is 72.36%. The confusion matrix obtained is shown in Figure 10.5.

10.4.5 Random Forest Classifier (RFC)

This is a supervised learning algorithm and the vital advantage concerning this algorithm is that it can prevail for both classifications together with regression problems. The divergence between the Random Forest (RF) algorithm and decision tree algorithm is that in RF the progression of finding the root node and intense of the feature nodes will run randomly [40–45]. An RFC with 30 estimators is used to get an accuracy of 76.31%. Obtained misperception matrix is shown in Figure 10.6.

10.4.6 Neural Network Implementation

This is a computational network model (neurons) that are massively interconnected, operate in parallel. The origins are inspired by the biological

	0	1
0	30	10
1	11	25

Figure 10.5 Obtained confusion matrix for Decision Tree Classifier was 72.36%.

	0	1
0	36	4
1	14	22

Figure 10.6 Obtained confusion matrix for Random Forest Classifier was 76.31%.

nervous system. The associations between elements to a great extent decide the system work. Therapeutic Determination utilizing artificial neural networks is at present an exceptionally dynamic research territory in prescription and it is trusted that it will be even more generally utilized as a part of the biomedical framework.

10.5 Data Preparation

The data training is an important part to define the best variables that contribute to defining the optimal solution. This entails the detailed knowledge of ANN. The first level is known to be the input layer and the last level is the output layer. The surplus layers of units between the input and output layer are called a hidden layer. A function can be implemented by modifying the values of the weights between elements [45–50].

10.5.1 Multi-Layer Perceptron Neural Network (MLPNN) Algorithm and Prediction

The structure of MLPNN with the layers are significant inputs to the prediction of heart disease that have been used with their weights. Here, as a training algorithm, MLPNN with Back-Propagation is being used for the prediction of heart disease. This is done in a unidirectional way. The input neurons of the first layer "[X1, X2, X3....Xu]" in a feed-forward neural network forward their output to the neurons of the second layer as "[V11, V21...]" from which hidden layer "[Y1, Y2,...Ym]" evaluates the weights "[W11, W22,......]" and gives the output layer as "[z1...]".

$$\Delta Wxz\_\{m\} \mid = \varepsilon\delta p\alpha X\_\{q\}$$

$$\partial z_P = \alpha z_{P(\,1\cdot -\alpha^{zP})}(t - az_P - \alpha z_p)$$

if output node.

$$\partial Zp = aZp\left(1\text{-}a\frac{n}{i=0}Zp\right)\Sigma aYiWzyi$$

if intermediate node.

Where x is the emitting node, z is the receiving node, y is the layer which follows z, W is weight, a is for activation, delta values with ε as the learning rate.

10.6 Results Obtained

The result obtained from the above experiment gives roughly 85 to 90% accuracy of heart disease. We have collected the data which was freely available on the internet. The age factors were mainly considered and trained data and tested data by using various deep learning techniques like RF, MLPNN, and NN. The architecture details of ANN is as shown in Table 10.2.

10.6.1 Accuracy

The obtained accuracy over 500 epochs is shown in Figure 10.7, where Training accuracy is 86.78% and Validation accuracy is 84.21%.

10.6.2 Data Analysis

The data analysis tested on various parameters was carried out for the cause of heart attack by considering various parameters and the graphs is as shown from Figures 10.8 through 10.21.

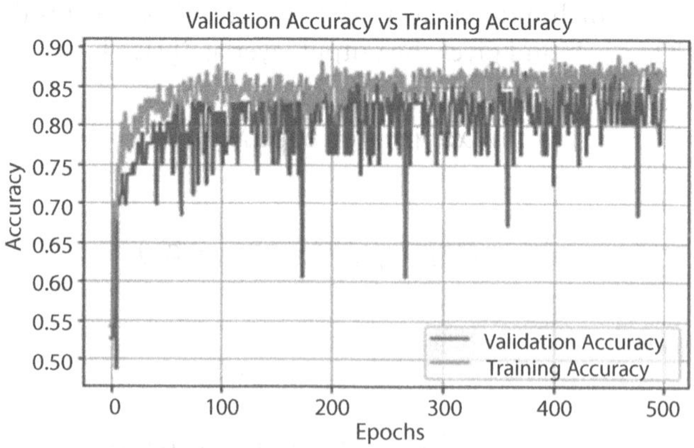

Figure 10.7 Validation accuracy v/s training accuracy.

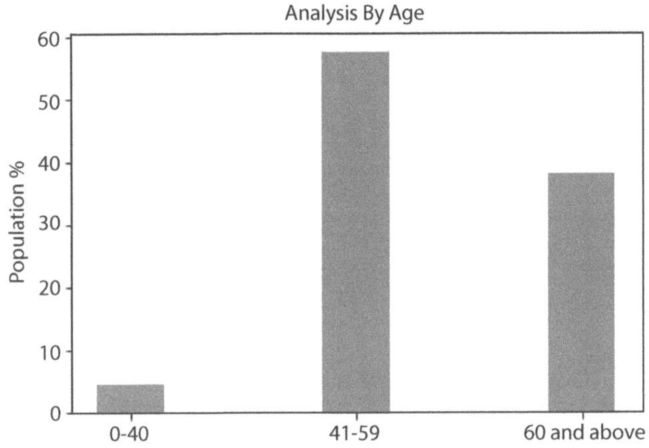

Figure 10.8 Analysis by age.

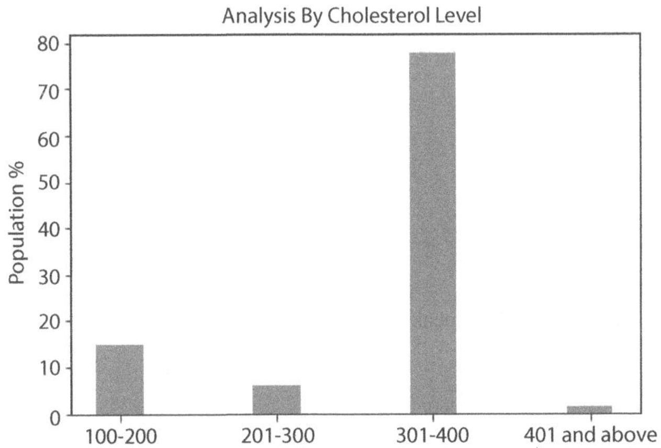

Figure 10.9 Analysis by cholesterol type.

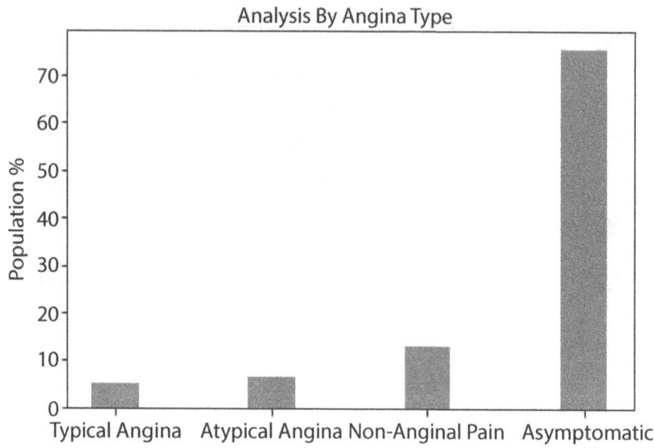

Figure 10.10 Analysis by chest pain.

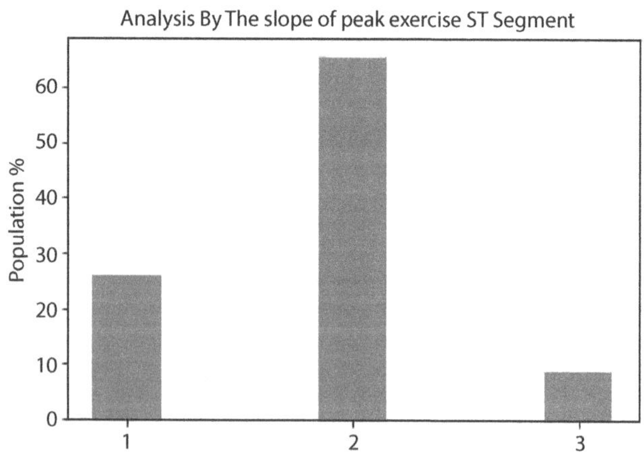

Figure 10.11 Analysis by slope peak.

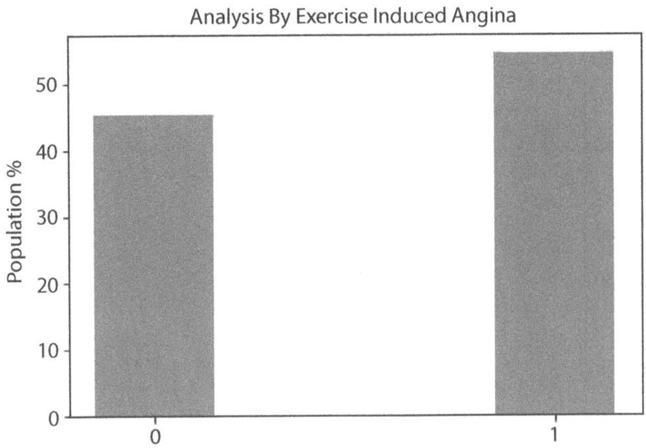

Figure 10.12 Analysis by Induced angina.

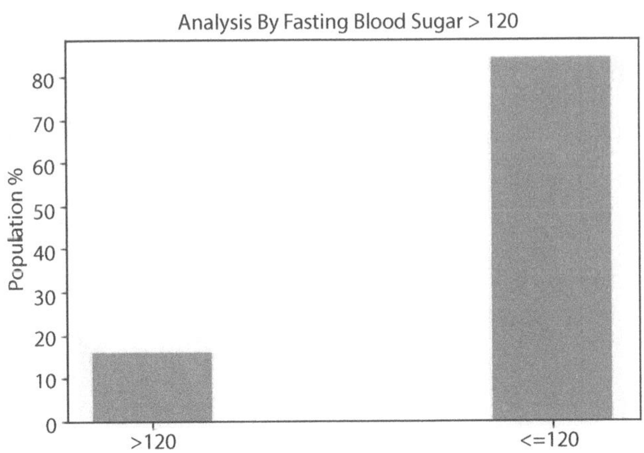

Figure 10.13 Analysis by Fasting Blood Sugar.

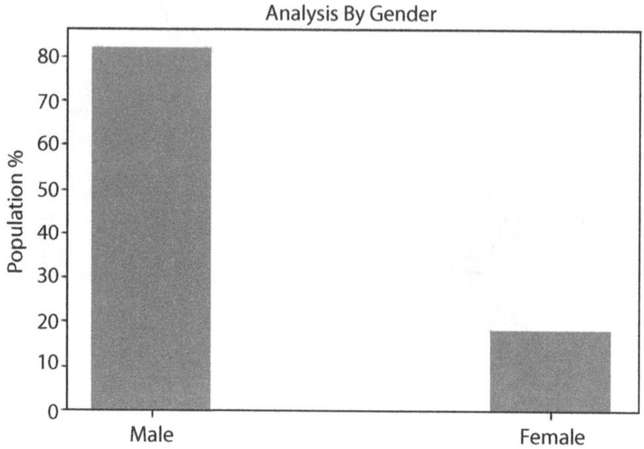

Figure 10.14 Analysis by Gender.

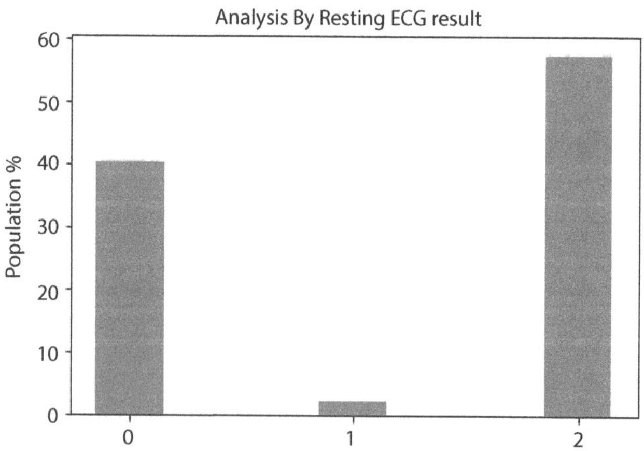

Figure 10.15 Analysis by Resting ECG.

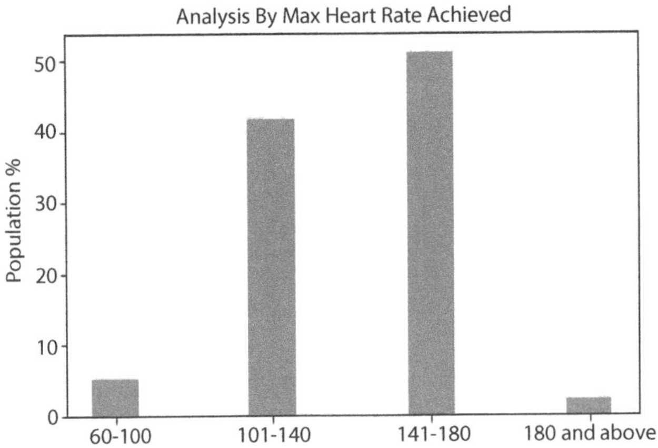

Figure 10.16 Analysis by Max Heart Rate.

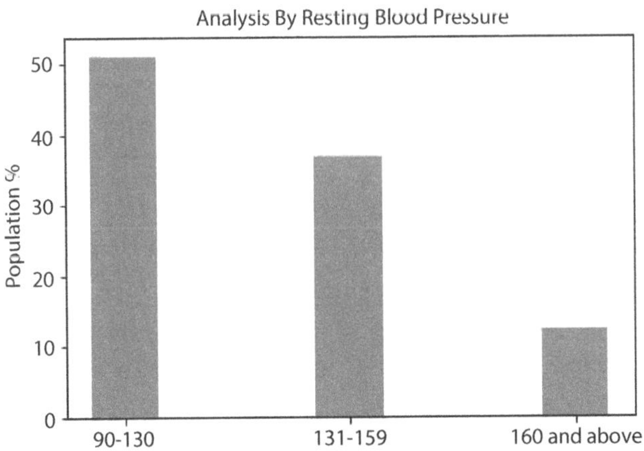

Figure 10.17 Analysis by Resting.

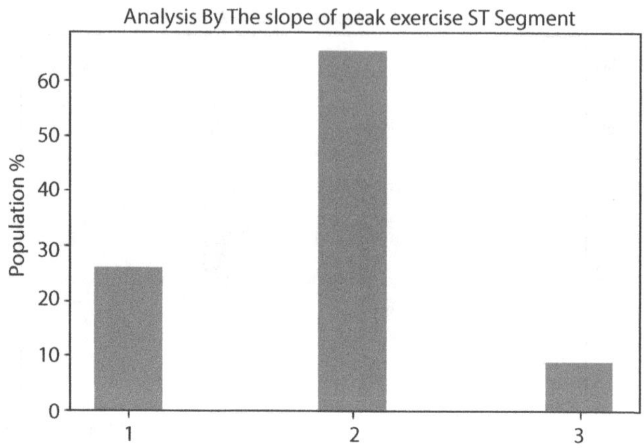

Figure 10.18 Analysis by slope peak.

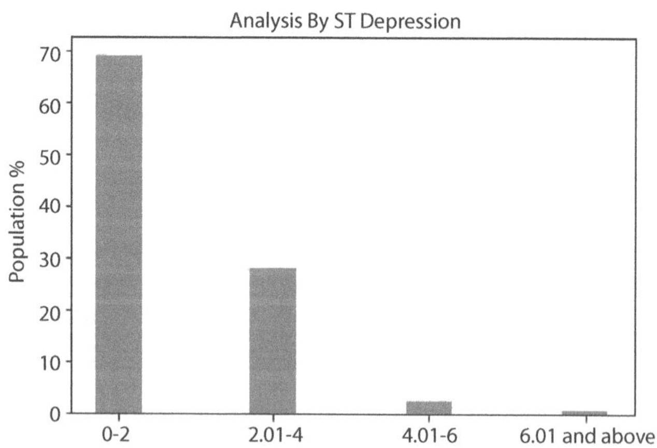

Figure 10.19 Analysis by Depression.

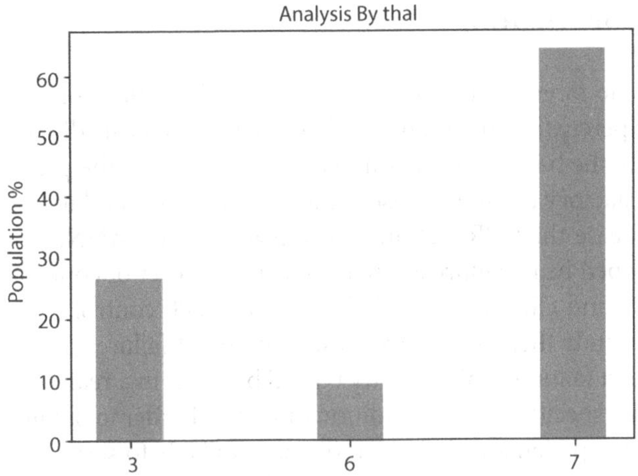

Figure 10.20 Analysis by Thalassemia.

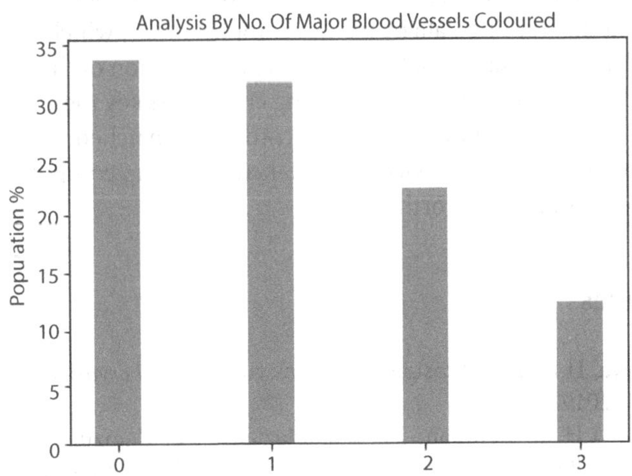

Figure 10.21 Analysis by Fluoroscopy.

10.7 Conclusion

The principle target of this work is to assess the utilization of ANN with Multilayer perceptron neural network for analysis of heart diseases. The performance of the back propagation neural network with the prediction accuracy is satisfactory but to increase the level of accuracy and to know the type of heart disease the CBR technique is integrated with ANN. Those patients who developed heart failure are matched against a set of control patients on age, gender, and clinics, and ended up with 45,000 control patients. Notice that in this study there are a fewer cases with heart failures than the controls without heart failures. This is pretty typical because, in a real-world scenario, patients with specific disease conditions are often harder to obtain while there are a lot more patients without that disease available to serve as controls. To summarize, in a case-control study, first, the cases are identified, then try to match them to a set of control patients. In the cohort study, all the patients exposed to the risk are identified, and the matching criteria are not involved. Evaluation of predicted models is one of the most crucial steps in the pipeline. Using some training samples, the idea is to devote the model, but test this training model on some other unseen samples, ideally from future data. It may be noted that the training error may not be very useful because it can easily overfit the training data by using a complex model which does not generalize well to future samples. For a better approximation of the performance of the model on future samples, the testing error is the key metric. It would be interesting for further work, if any old record is not matched with the present record then expert doctor advice can be taken by certain computing methods like mobile, cloud, and so forth.

References

1. Da Silva, H.P. *et al.*, Biosignals for Everyone. *IEEE Pervasive Comput.*, 7, 4, 64–71, 2019.
2. Majid, A.H.A., Rahman, M.A.A., Mazlan, S.A., Zamzuri, H., Biosignals based intelligent control interface for current-induced physiological devices, in: *2015 10th Asian Control Conference (ASCC)*, 2015, May, IEEE, pp. 1–5.
3. Baumert, M., Porta, A., Cichocki, A., Biomedical signal processing: From a conceptual framework to clinical applications [Scanning the Issue]. *Proc. IEEE*, 104, 2, 220–222, Feb. 2016, doi: 10.1109/JPROC.2015.2511359.
4. Vijayashree, J. and SrimanNarayanaIyengar, N.C., Heart disease prediction system using data mining and hybrid intelligent techniques: A review. *Int. J. Bio-Sci. Bio-Technol.*, 8, 4, 139–148, 2016.

5. Zhu, Y., Wu, J., Fang, Y., Study on application of SVM in prediction of coronary heart disease. *Sheng wu yi xue gong cheng xue za zhi= J. Biomed. Eng.= Shengwu yixue gongchengxue zazhi, 30*, 6, 1180–1185, 2013.
6. Soni, J., Ansari, U., Sharma, D., Soni, S., Predictive data mining for medical diagnosis: An overview of heart disease prediction. *Int. J. Comput. Appl., 17*, 8, 43–48, 2011.
7. Austin, P.C., Tu, J.V., Ho, J.E., Levy, D., Lee, D.S., Using methods from the data-mining and machine-learning literature for disease classification and prediction: a case study examining classification of heart failure subtypes. *J. Clin. Epidemiol., 66*, 4, 398–407, 2013.
8. Chowdhury, D.R., Chatterjee, M., Samanta, R.K., An artificial neural network model for neonatal disease diagnosis. *Int. J. Artif. Intell. Expert Syst. (IJAE), 2*, 3, 96–106, 2011.
9. Kadi, I., Idri, A., Fernandez-Aleman, J.L., Knowledge discovery in cardiology: A systematic literature review. *Int. J. Med. Inf., 97*, 12–32, 2017.
10. Han, S.H., Kim, K.O., Cha, E.J., Kim, K.A., Shon, H.S., System framework for cardiovascular disease prediction based on big data technology. *Symmetry, 9*, 12, 293, 2017.
11. Farran, B., Channanath, A.M., Behbehani, K., *et al.*, Predictive models to assess risk of type 2 diabetes, hypertension and comorbidity: Machine-learning algorithms and validation using national health data from Kuwait—a cohort study. *BMJ Open*, 3, e002457, 2013. doi: 10.1136/bmjopen-2012-002457
12. Heydari, M., Teimouri, M., Heshmati, Z., Alavinia, S.M., Comparison of various classification algorithms in the diagnosis of type 2 diabetes in Iran. *Int. J. Diabetes Dev. Ctries, 36*, 2, 167–173, 2016.
13. Colubri, A., Silver, T., Fradet, T., Retzepi, K., Fry, B., Sabeti, P., Transforming clinical data into actionable prognosis models: Machine-learning framework and field-deployable app to predict outcome of Ebola patients. *PloS Negl. Trop. Dis., 10*, 3, e0004549, 2016.
14. Jabeen, A., Ahmad, N., Raza, K., Machine learning-based state-of-the-art methods for the classification of rna-seq data, in: *Classification in BioApps*, pp. 133–172, Springer, Cham, 2018.
15. Ahmed, S.S., Dey, N., Ashour, A.S., Sifaki-Pistolla, D., Bălas-Timar, D., Balas, V.E., Tavares, J.M.R., Effect of fuzzy partitioning in Crohn's disease classification: A neuro-fuzzy-based approach. *Med. Biol. Eng. Comput., 55*, 1, 101–115, 2017.
16. Azzabi, O., Njima, C.B., Messaoud, H., New approach of diagnosis by timed automata. *Int. J. Ambient Comput. Intell. (IJACI), 8*, 3, 76–93, 2017.
17. Khachane, M.Y., Organ-based medical image classification using support vector machine. *Int. J. Synth. Emot. (IJSE), 8*, 1, 18–30, 2017.
18. Dey, N. and Ashour, A. (Eds.), *Classification and clustering in biomedical signal processing*. p. 463, IGI Global, Pennsylvania, USA, April 2016, DOI: 10.4018/978-1-5225-0140-4.

19. Touati, F. and Tabish, R., U-healthcare system: State-of-the-art review and challenges. *J. Med. Syst.*, *37*, 3, 1–20, 2013.
20. Moridani, M.K., Setarehdan, S.K., Nasrabadi, A.M., Hajinasrollah, E., Analysis of heart rate variability as a predictor of mortality in cardiovascular patients of intensive care unit. *Biocybern. Biomed. Eng.*, *35*, 4, 217–226, 2015.
21. Zhu, Y., Wu, J., Fang, Y., Study on application of SVM in prediction of coronary heart disease. *Sheng wu yi xue gong cheng xue za zhi= J. Biomed. Eng.= Shengwu yixue gongchengxue zazhi*, *30*, 6, 1180–1185, 2013.
22. Soni, J., Ansari, U., Sharma, D., Soni, S., Predictive data mining for medical diagnosis: An overview of heart disease prediction. *Int. J. Comput. Appl.*, *17*, 8, 43–48, 2011.
23. Kadi, I., Idri, A., Fernandez-Aleman, J.L., Knowledge discovery in cardiology: A systematic literature review. *Int. J. Med. Inf.*, *97*, 12–32, 2017.
24. Tu, J., Donovan, L., Lee, D., *Quality of Cardiac Care in Ontario: EFFECT (Enhanced Feedback for Effective Cardiac Treatment)*, Institute for Clinical Evaluative Sciences, Toronto, Canada, 2004.
25. Tu, J.V., Donovan, L.R., Lee, D.S., Wang, J.T., Austin, P.C., Alter, D.A., Ko, D.T., Effectiveness of public report cards for improving the quality of cardiac care: The EFFECT study: a randomized trial. *Jama*, *302*, 21, 2330–2337, 2009.
26. Santhanam, T. and Ephzibah, E.P., Heart disease prediction using hybrid genetic fuzzy model. *Indian J. Sci. Technol.*, *8*, 9, 797, 2015.
27. Poornima, V. and Gladis, D., A novel approach for diagnosing heart disease with hybrid classifier. *Biomed. Res.*, *29*, 11, 2018.
28. Dwivedi, A.K., Performance evaluation of different machine learning techniques for prediction of heart disease. *Neural Comput. Appl.*, *29*, 10, 685–693, 2018.
29. Zhang, C. and Ma, Y. (Eds.), *Ensemble Machine Learning: Methods and Applications*. pp. VIII, 332, Springer Science & Business Media, Springer-Verlag New York, 2012.
30. Dietterich, T.G., Ensemble learning, in: *The Handbook of Brain Theory and Neural Networks*, 2, pp. 110–125, 2002.
31. Breiman, L., Bagging predictors. *Mach. Learn.*, *24*, 2, 123–140, 1996.
32. Skurichina, M. and Duin, R.P., Bagging, boosting and the random subspace method for linear classifiers. *Pattern Anal. Appl.*, *5*, 2, 121–135, 2002.
33. Ho, T.K., The random subspace method for constructing decision forests. *IEEE Trans. Pattern Anal. Mach. Intell.*, *20*, 8, 832–844, 1998.
34. Detrano, R., Janosi, A., Steinbrunn, W., Pfisterer, M., Schmid, J.J., Sandhu, S., Guppy, K.H., Lee, S., Froelicher, V., International application of a new probability algorithm for the diagnosis of coronary artery disease. *Am. J. Cardiol.*, *64*, 5, 304–310, 1989.

35. SPSS, *IBM SPSS Statistics for Windows, Version 20.0*, IBM SPSS Statistics for Windows, Version 24.0., IBM Corp, Armonk, NY, Released 2016. https://www-01.ibm.com/support/docview.wss?uid=swg21476197
36. Groves, P., Kayyali, B., Knott, D., Kuiken, S.V., The 'big data' revolution in healthcare: Accelerating value and innovation. © 1996-2021 McKinsey & Company, New York, 2016.
37. Langdridge, D., Recovery from heart attack, biomedicalization, and the production of a contingent health citizenship. *Qual. Health Res.*, *27*, 9, 1391–1401, 2017.
38. Lee, S.H., Kim, M., Chung, H., Yang, S., Designing of efficient technique blocking abnormal packets through correlation analysis in the healthcare environment. *Indian J. Sci. Technol.*, *9*, 24, 1–5, 2016.
39. Ebenezer, J.G.A. and Durga, S., Big data analytics in healthcare: A survey. *Arpn J. Eng. Appl. Sci.*, 10, 3645–3650, 2015.
40. Ciccone, M.M., Aquilino, A., Cortese, F., Scicchitano, P., Sassara, M., Mola, E., Rollo, R., Caldarola, P., Giorgino, F., Pomo, V., Bux, F., Feasibility and effectiveness of a disease and care management model in the primary healthcare system for patients with heart failure and diabetes (Project Leonardo). *Vasc. Health Risk Manage.*, *6*, 297, 2010.
41. Lee, B. and Jeong, E., A design of a patient-customized healthcare system based on the hadoop with text mining (PHSHT) for an efficient disease management and prediction. *Int. J. Software Eng. Appl.*, *8*, 8, 131–150, 2014.
42. Fard, A.D. and Mojtabaei, M., The effect of cognitive behavioral stress management and psychological well-being and adherence to treatment in patients with coronary heart disease (Chd). *Int. J. Humanit. Cult. Stud. (IJHCS)*, *2*, 2015.
43. Kumar, K.G., Arvind, R., Keerthan, P.B., Kumar, S.A., Dass, P.A., Wireless methodology of heart attack detection. *Int. J. Sci. Res. Dev.*, *2*, 673–676, 2014.
44. Hsieh, J.C., Li, A.H., Yang, C.C., Mobile, cloud, and big data computing: Contributions, challenges, and new directions in telecardiology. *Int. J. Environ. Res. Public Health*, *10*, 11, 6131–6153, 2013.
45. Prerana, T.H.M., Shivaprakash, N.C., Swetha, N., Prediction of heart disease using machine learning algorithms-Naïve Bayes, introduction to PAC algorithm, comparison of algorithms and HDPS. *Int. J. Sci. Eng.*, *3*, 2, 90–9, 2015.
46. Lee, V. Y., Tan, S. C., Chee, J. K., Idea Identification and Analysis (I2A): A search for sustainable promising ideas within knowledge-building discourse, in: *Transforming Learning, Empowering Learners: The International Conference of the Learning Sciences (ICLS) 2016*, C. K. Looi, J. L. Polman, U. Cress, P. Reimann, (Eds.). vol. 1, International Society of the Learning Sciences, Singapore, 2016.
47. *The Use of Big Data in Public Health Policy and Research: Background information document*, pp. 1–17, European Commission Directorate-General for Health and Consumers, Unit D3 eHealth and Health Technology Assessment, Brussels, 29 August 2014.

48. Ghadge, P., Girme, V., Kokane, K., Deshmukh, P., Intelligent heart attack prediction system using big data. *Int. J. Recent Res. Math. Comput. Sci. Inf. Technol.*, *2*, 2, 73–77, 2015.
49. Winter, R., Big data: Business opportunities, requirements and oracle's approach. p. 2, Executive Report, Winter Corporation, Tyngsboro, Massachusetts, United States, 2011.
50. Russom, P., Big Data Analytics, TDWI Best Practices Report, TDWI Research, 1105 Media Inc., Woodland Hills, CA, Fourth Quarter 2011, http://tdwi.org/research/2011/09/best-practices-report-q4-big-data-analytics/asset. aspx

11

Supervising Healthcare Schemes Using Machine Learning in Breast Cancer and Internet of Things (SHSMLIoT)

Monika Lamba[1]*, Geetika Munjal[2†] and Yogita Gigras[1‡]

[1]The NorthCap University, Gurugram, India
[2]Amity University, Noida, India

Abstract

The 21st-century era of advancement in Information Technology and emerging applications like the Internet of Things (IoT) have altered the lifestyle of people and its contribution in the field of healthcare is commendable. The information related to the body and mental strength can help in bringing encouraging transformation in the healthcare industry with the usage of networked sensors. The data collected with the help of IoTs can be evaluated using machine learning processes and professional to do advanced research and provide suitable recommendations. These scenarios already exit but proper research and development needs to be carried out of the historical data for predicting upcoming troubles utilizing strict analytics. The present document is based on different machine learning applications and knowledge is extracted for real time analysis of data and secluded health supervision built on IoT infrastructure. The system formulated will provide recommendations based on past and pragmatic data stored in healthcare. The author has formulated a structure to expose acquaintance in breast cancer which will help in reliable and authentic decision making. This paper has estimated a forecasting scheme for breast cancer using machine learning technique like SVM, Bayes Net, J48, MLP, Decision Trees, etc. in experimental results.

***Keywords*:** IoT, machine learning, healthcare, sensor, breast cancer

**Corresponding author*: missmonikalamba@gmail.com
†*Corresponding author*: munjal.geetika@gmail.com
‡*Corresponding author*: yogitagigras@ncuindia.edu

Kavita Sharma, Yogita Gigras, Vishnu Sharma, D. Jude Hemanth and Ramesh Chandra Poonia (eds.) Internet of Healthcare Things: Machine Learning for Security and Privacy, (241–264) © 2022 Scrivener Publishing LLC

11.1 Introduction

Breast Cancer (BC) represents a truly devastating illness amidst all the illnesses in the medical field. It represents one of the important explanations of mortality among the women all over the globe and is one of the critical common diseases which significantly affect ladies as per AICR (American Institute for Cancer Research) [1]. In 2018, the count of BC cases was 2 million. BC is the 5th bigger cause of death in comparison to other types of cancer. That is why the early discovery of this disease may decrease the death rates. Cells of BC develop unnaturally and steadily enhance the influenced cell rate, leading to BC. It is a cancerous tumor produced and built-in cells of the breast. A band of dividing cells make an extra tissue known as tumors, plus such cancers are categorized as benign i.e., non-cancerous and malignant as cancerous. With the development of modern technology, in various countries, the 5-year SR (survival rate) in the initial phase is about approx. 90% and falls at the initial stage to 24% for identification of BC in medical science [2]. For the prediction and diagnosis of BC, multiple invoice-based methods are used. Biopsy [3] is one of the methods resulting in high and good accuracy with the help of breast tissues and taking biopsy is too painful for the patients. A mammogram is another method [4] considered for BC diagnosis. In a mammogram, a 2-D image of the breast is specially designed, but this method does not perform well in diagnosing the benign type of cancer successfully. MRI (magnetic reasoning imaging) [5] is an invoice-based method; it is a highly complex test and delivers the best results for 3-D images. More time is needed in all these methods to generate the output [6]. To help the problem of complexities in the three above mentioned methods for BC diagnosis, mainly machine learning (ML) methods i.e., non-invasive are more trustworthy and effective. In literature, machine learning methods are used to categorize BC as either malignant or benign. The detailed literature of ML has been described for the detection and diagnosis of BC.

Machine Learning [7], being the branch of AI (artificial Intelligence) performs a vital starring role in the detection and prediction of innumerable disease like breast cancer as data in the medical term have been digitalized. In the preceding decade, research experts tried to discover and study a variety of ML practices in the approach of health attention. The approach follows steps of the first recognition of anomalies and second classifying the anomalies into the relevant category. Both the steps require supervised taxonomy job, initially linguistic knowledge by categorizing for each pixel as it fit into questionable contusion and lastly permitting for additional investigation causing inappropriate classification. Internet of Things (IoT) [8] is a kind of legitimate modernization enabled network. With the help of IoT, the whole thing has been correlated with the web, suggesting brand-new possibilities to enrich pursuits crossways throughout farming, manufacturing, and the economy. Out of everything, healthcare services symbolize one of the essential efficient and collective obstacles that every single people will face. Human services experts, therapists, entrepreneurs, and the rest wellbeing specialists are a tough burden for accepting emerging obligations from the secretive class and widespread high society. As the basic quality of livelihood of a human being like impaired, senior citizen and person experiencing from never concluding illness residing in their homes and remote areas are escalating exponentially. Therefore, the prerequisite of isolated patients noticing is transforming into a typical procedure. At present, the assessment is how to deliver suitably and excellent supervision beyond the restorative facility situation. In this framework, advanced automated therapeutic maintenance executives are expected to receive reachable through a structure at any time, everywhere and to anyone. Instead of this, IoT possesses as a proposed structure for recuperative sensor communication foundation which is going to render the utilization of remote upgrades that energize which is uninterrupted communication of knowledge regarding a patient's disease to healthcare professionals. Numerous useful devices are easy to use that can differentiate a variety of medical circumstances—blood pressure, breathe, heart rate, etc. since a consumer's touch and switch such accounts through cell phones. With the support of connected instruments to use the mobile devices there is a want to adapt endless inspection of the sufferer's health condition and dealing with their medication from isolated regions. In this way, IoT will presume a remarkable involvement in the coming-of-age healthcare services supplying. Nevertheless, the enhancement and implementation of the healthcare system via cell phones display a couple of obstacles, several of them being brilliant data gathering as well as interoperable, ease of access of diverse resources, security, safety, and data inconsistency. Cloud computing provides an opportunity for the service to get to the shared resources and fundamental groundwork on a single individual and simply, providing services by the petition throughout the system and accomplishing responsibilities specified by the advancing demands. IoT in accordance with the medical treatment surveillance framework comprises "internet of security sensor events". These are things that establish enormous quantities of data that may not be considered by a physician. The physician's underlying worry is the reality that he ought to settle down on adoptions concerning patient's health for whom he wishes to extract the data over a particular patient since the deluge of medical records equivalent to the enormous count of patients. With the assistance of IoT operator, the health information was applied by transmitting it to the cloud and where cloud assistance to the amount of data and offer appropriateness to the large data analysis.

Endless frightening of the individual patient's wellbeing is authorized by using data analysis and mining. Individual patient's data like body pressure, blood temperature etc. will be collected using the various body sensor connections. Additionally, the information collected is transmitted to the Multimedia Tools and Applications via IoT devices like mobile on a cloud-based network. With the support of a cloud-based network, the information will be handled and kept efficiently. Physicians act remotely through data handling and the ability to upload their replies onto the cloud. A major business project along with the medical care project represents the highest excellence of assistance available for detecting illness expertise and executing a facility of valuable therapies for the patients. Judgment has been difficult and complicated but critical enterprise must be executed efficiently. The forecast is rendered regularly based upon the doctor's knowledge and understanding which may lead to inaccurate and adverse reactions. Consequently, it appears as a requirement of a computerized medical evaluation scheme to devise which can take advantage of the collected information and the judgment assistance system. Such a scheme could help in diagnosing the illness more frequently with reduced medical examinations and major symptoms that are understood by the sick person. For the implementation of this necessity, creators are making use of the healthcare data and information technology administered by clinics. They accumulated huge quantities of data to extract concealed information and construct a smart disease forecast for the system using a historical databank of numerous diseases [9]. Consequently, ML methods are authorized for handling a large quantity of patient information that was previously hidden and significantly useful data in an applicable manner throughout the e-healthcare system [10].

The objective of the present chapter is to evaluate the efficacy of current ML methods aimed at different community dataset for medical treatment products and to introduce an effective methodology in forecasting the illness depending upon the IoT network. The chapter is summarized as follows. Section 11.2 identifies the linked work carried out in the healthcare for illnesses of the prophecy. Section 11.3 exemplifies research materials and techniques; Section 11.4 explains IoT and disease. Section 11.5 portrays experimental outcomes and Section 11.6 concludes the paper.

11.2 Related Work

The expertise of the healthcare field will be building on the experience or the proof amassed through medical specialists. A person's body is like a complex piece of equipment that comprises several parts and components that may be influenced by numerous reasons. Consequently, the simulation of the problem dis-functionality or functionality is a monotonous assignment. Therapeutic statistics could be used for making reflections, observation, and expectations to the definitive diagnosis of a variety of diseases. Machine-learning techniques offer good quality layout such as tools that might learn from preceding knowledge and set up predictions that extremely beneficial in e-health applications.

Hameed has introduced the e-healthcare provision structure that is based upon cloud computing in Iraq and service concerned with architecture. The planned outline varies on SOA (service-oriented architecture) suggested dissimilar departments alike improving outlay the executive's time, provisions patient's summary and takings the truthful dedicated choice [11]. Hsu *et al.* suggested a paradigm centered on fitness information necessary for the BC risk evaluation. Throughout this dimension reduction and sampling methods have been requested to pre-procedure the assessment data. Later, a variety of taxonomies were intended for the risk forecast [12]. Parekh *et al.* have introduced a cloud-centered structure for the healthcare scheme by utilizing gathering and grouping as a data mining method. Such an approach authorized sufferers to obtain the necessary for the wellbeing presidencies all through a flexible implementation by just a few ticks [13]. SVM and Naïve Bayes techniques are used by Vijayarani and Dhayanand to do the analysis and forecasting for kidney illness. The overall analysis was based on execution time and accuracy. With the support of results, SVM showed better in comparison with Naïve Bayes [14]. Thyroid disease forecasting by Turanoglu-Bekar *et al.*, by using decision tree mining methods like LADTree, Random Forest, LMT, etc. Evaluation of performance measure was on six parameters including MAE, FME, accuracy, PRE, kappa statistics and REC. The results show the best performance in the case of NBTree [15]. K-means and particle swarm intelligence were used by Verma *et al.* [16], to predict the risk associated with coronary artery disease. They utilized a variety of learning algorithms like fuzzy unordered rule induction, Multinomial Logistic Regression (MLR), C4.5, and Multi-Layer Perceptron (MLP) for obtaining data events. Data collection was done from the Department of Cardiology, Shimla, India, and the dataset comprises 335 instances and 26 features. MLR attained a maximum accuracy of 88.4%. Usage of MLP (Auto MLP) for forecasting diabetes was designed by Jahangir *et al.* [17] using class outlier detection in the pre-processing step followed by automatically turning the parameters in the training process. A forecasting model for disease avoidance and everyday surveillance called ViSiBiD was proposed by Forkan *et al.* [18], for the motive of important sign of the sick person. The ML methods in conjunction with certain Map-Reduce enactment were being utilized for understanding the cloud program. Making use of 4,893 patients' data set that is available to the public, six bio-signals were found to diverge from various characteristics to watch the information activities. Information contained in events is gathered in periods of 1–2 h. Results demonstrate random forest possesses the absolute best precision of 95.85% when contrasted to the other approaches. A technique for diagnosis diabetics proposed by Osman and Aljahdali [19] based upon enhanced feature extraction. For categorization, K-means and SVM techniques were used. Its experimental assessment illustrates that the planned method requires a better categorization rate of diabetes prediction than the already existing one. A unique statistical learning theory for cancer forecasting is suggested by Zheng *et al.* [20] and the method was accomplished for multi and binary class troubles. The SVM method generates the larger hyperplanes in high-level spatial space, which expands the distance among the data points, multimedia equipment and applications with support vectors are utilized to generate hyper-plane. The SVM offers improved accuracy although expensive by the conditions of computational time. Pavleen *et al.* [21] presented a system for healthcare depending on IOT and Random forest. The designed model improves by interaction among doctor and patients. Utilizing multiple datasets concerning numerous diseases resembling heart disease, BC, liver disorder, thyroid for the assessment purpose to enhance the efficacy of the model. Using ML methods like SVM, Decision tree, MLP, K-NN, the highest accuracy of 97.26% was accomplished on the Dermatology with the support of Random Forest. Moreover, the security of the data is defined as one of the vital IOT based devices as it collects information using available technology and transmitting the data to other IoT devices. Using the help of ML classifier SVM and recursive feature selection to enhance the performance classification is done by Menon *et al.* [22], achieved accuracy as 99%, sensitivity as 98%, MCC as 99% and specificity as 99%. The dataset used the Wisconsin Diagnostic BC and dividing the dataset into testing and training module for better prediction. The healing and therapy are going to be more efficient for BC. Connection with the application of the suggested scheme is extremely trustworthy in all facets of IoT medical care for BC. By the literature assessment and analysis, it should be noted that the information about the utilization of meta-heuristic procedures stands omitted by most of the previously prevailing researchers to progress the classification percentage and ranking of existing machine learning practices.

In numerous studies of microarray, classifiers have already been built depending upon that gene signatures to accurately forecast medical outcomes for many different cancer sufferers. Still, the signature is derived from the various studies that frequently suffer from inadequate toughness when they are used in the category/ranking of information from the autonomous sets from which they have been created. Zhang [23] introduced an unsupervised feature learning structure by incorporating a PCA (principal component analysis) algorithm and autoencoder neural network to recognize various attributes since gene manifestation profiles. Throughout the experiments establishing an alternative classifier with an identical classifier learning approach to accomplish as a reference point to the suggested method, wherever the only discrepancy is the training inputs. Experimental findings show that the recommended method utilizing deep learning methods is performing better than others.

If BC is diagnosed by the side of an early phase, it can be one of the curable types of cancer. Thanks to the development of the discovery expertise, an ever-increasing quantity of clinical information and high-dimensional characteristics may be used for BC to be diagnosed. The single most high-dimensional information that can be found has remained contributing to advances in analytical technology, although additionally, it receives a massive amount of data-processing redundancy. Extricating critical information and a reduction in the feature is essential to effective forecasting and an exact treatment choice. The earlier work for BC being diagnosed has been primarily based on labelled data that is hard to acquire. To solve this problem, Milon *et al.* demonstrated a different plan, that incorporates deep learning by the unsupervised feature extraction process, for BC with SVM and stacked autoencoder. SVM is utilized to categorize the samples with brand-new functionality into benign or malignant tumors. The suggested procedure was analyzed on the WDBC (Wisconsin Diagnostic Breast Cancer). Performance is assessed applying different methods and in comparison, including the earlier published results. A contrast between the findings indicates that the proposed SAE-SVM approach increases the accuracy by 98.25% and outstrips the additional methods. Deep learning based on unsupervised feature extraction dramatically enhances the efficiency of classification and delivers an encouraging approach to BC diagnosis [24].

Islam [25], introduce a narrative sensory system for the estimate of BC and introduces SVM and K-NN that make up the supervised ML methods for BC detection through its characteristics. The overall accomplishment of the suggested scheme is evaluated by contemplating accuracy, MCC, false discovery rate, sensitivity, specificity, and untrue omission rate. The tactic provides the best outcome equally for testing and training. Additionally, the methods attained the accuracy of 98.57% by SVM and 97.14% by K-NN independently in conjunction with the specificity of 95.65 and 92.31% in the testing phase.

Image-processing methods for using ML are used extensively in the field of the health domain to enhance the initial recognition of cancer-causing tumors in the breast. Chaitanya *et al.*, through a suggested method, supervised learning methods that are being applied to obtain cancer to define features and categorize cancerous images from the usual mammogram images. The supervised system is originally qualified with extricating 13 features to each one from the 30 images dataset. The extricated features of the image by the test are linked to the features extracted from the database images to identify and forecast the cancer tumor in the image. SVM and KNN are intended for categorization. Centered upon the analysis, the system can provide a classification with the precision of 95% with SVM and 97% with KNN. Additionally, a user-friendly chatbot is established by utilizing Dialog Flow, which cooperates with the patients to forecast cancer depending upon the signs recognized by the patient. This chatbot may be applied by the patient to distinguish if the indications are prone to cancer [26].

The truthful and cost-effective identification of BC is especially crucial for healing and medication at the beginning stages throughout the IoT medical environment. IoT observed the change in life for the past several years that offers an approach to investigate both the past and real-time data through the developing role of data mining (DM) and artificial intelligence (AI). The present requirement does not successfully analyze the BC during the initial stages. Consequently, the initial diagnosis of BC substantially stands a major question meant for investigators and medical professionals. As per the doctor's opinion, regular self-analysis of the breast is very helpful to identify signs at the early stage. Although today people are well educated about the revolution in technology such as mobile phones, smart devices and preoccupied in their tedious works. People are expecting economical and non-ionic technological help for breast health surveillance. Kavitha [27] presented an innovative model named ICHC (IoT-cloud-based healthcare) to discover BC sign at the quickest and earliest. With the help of sensors, mental and physical health data could be used to bring a beneficial transformation for the persons who do not have immediate contact to health surveillance scheme.

Azar [28] introduced an approach for BC diagnosis using RBF (radial basis function), PNN (probabilistic neural network), and MLP (multi-layer perceptron). Three classifiers were trained and tested. Performance was presented using sensitivity, accuracy, and specificity. The accuracy achieved was 97.8 % for training and 97.66% for testing obtained by MLP. Alickovi *et al.* [29] proposed a prediction method using a feature selection method called GA (genetic algorithm) and for classification (RF) rotation forest. The accuracy achieved was 99.48% with RF using GA for the selection of features. Ahmad [30] presented a system called GA-MOO-NN for BC detection. GA was used for selecting relevant features. Dataset is divided into three parts: a) one for training i.e., 50% b) second for testing i.e., 25% and c) third for validation i.e., 25%. Resulted obtained are 98.85% as best accuracy and 98.1% as average accuracy. Hasan [31] presented a method using multi-gene genetic programming. Accuracy of 98.28% was obtained using 10-fold cross-validation. Albrecht [32] introduced a method used for the diagnosis of BC by securing accuracy of 98.8%. PenaReyes [33] presented a method called fuzzy-GA and secured an accuracy of 97.36%. Akay [34] introduced a tactic using F-score for selecting features and SVM for classification. Zheng [35] presented a method using k-means for selecting a feature and extracting together with SVM. This technique secured maximum accuracy and low cost of computation. Madevi presented PCA (hybridized principal component analysis) along with various classifiers implemented on BC datasets [36]. The author introduced a tactic for the diagnosis of BC using Pareto ANN (artificial neural network). The results based on the experiment shows proposed method achieved low computation cost and good accuracy [37]. Marcano [38] introduced a method that obtained 99.26% accuracy using artificial meta plasticity MLP. A method based on a decision tree and using undersampling to stabilize the data [39]. Author Onan created a fuzzy-rough for feature selection by uniformity [40]. Sheikhpour *et al.* [41] proposed a tactic depending on PSO (particle swarm optimization) together with non-parametric kernel density assessment. Rasti *et al.* [42] diagnosed BC by achieving 96.39% accuracy and the method used was CNN (convolutional neural networks). Ani introduced an IoT depending on diagnostic and monitoring tool with the help of an ensemble classifier with the result of accuracy as 93% [43]. An IOT based ECG monitoring system was planned by Yang [44] for intelligent healthcare and showed good performance in diagnosing BC. Machine learning tactics are used along with REF (recursive feature selection) to enhance the performance of SVM. The work also used technique like LLBF [45], MRMR, BFO [46], relief [47], LASSO and feature selection in two stages [48]. The testing or training data was a dividing validation method to choose the best hyperparameters. The performance of the proposed model was judged depending on sensitivity, F1-score, MCC, specificity and accuracy. Datasets were collected from the UCI repository.

The critical contribution of the research are as follows:

a) Need for diagnosing BC at the early stage;
b) Importance of BC detection with the support of IoT;
c) Using different ML technique to classify the BC.

At last, concluding the successful detection of BC can be straightforwardly incorporated in the system of healthcare.

11.3 IoT and Disease

Healthcare research creates a variety of fields so that scientists in computer science, medicine, biomedical engineering, big data and microbiology discover themselves employed and cooperating on associated projects. To improve the health of patients, doctors work on diagnostic of multi-level and studies of laboratory [49]. To establish innovative medicinal and diagnostic tactics, biomedical engineers make use of biosensors to plan new medical tools. Using AI and computer system, scientists' efforts enhance the behavior of disease and envisage infections depending on symptoms. In the meantime, data scientists oversee medical research on treatment for illness like cancer and cooperating with clinics and hospital to assemble data on geolocations, medical care and other associated fields. In all the areas of expertise, it develops visibility that numerous gaps persist among them; these gaps represent foremost technical disputes in the direction of advancement of an extremely cohesive adaptable structure for healthcare. The straightforward way to establish this structure is to build an IoT depending on the cyberphysical key, to enable innovations and advances in all the areas. Enhancements in the IoT aid to establish meaningful enhancement in healthcare. For illustration, tactics like wearable sensors and chips of microfluid help to develop medical diagnostic in different variety of applications varying from laboratory to the hospital. In the future, IoT based devices will permit health givers to consistently assist people suffering from breast cancer, lung cancer, colorectal cancers and achieve a facet of basic care. It will help to deliver physician, the data and information to build genuinely data-driven plans of therapy, notably enhancing the chance of a fruitful revival. E-health (IOT-based healthcare) will be able to move the illness research ahead so that more precise prediction point of care and pace up the advancement of helpful pharmaceuticals. This needs to be brought into consideration that IOT e-health is not simply the straightforward stack of various worlds. As an alternative, these fragments are connected to the network jointly to adapt, predict, and assess in close together in real-time.

The concept of such a scheme is as mentioned:

a) E-health will vigorously process inquiries regarding patients at the genetic material (DNA), organ levels (kidney) and cellular (cells of blood), in a broad array of wearable biosensors. This schema will permit a large count of IoT based collective experiments on the way to real time, as more and more researchers and labs work in partnership to share data, stipulate reciprocated guidance, and influence this collectively shared database to notify verdicts on follow-up procedures and practices in the biochemical domain.
b) It will build and improve the accurateness of healthcare choice-making, by making use of the big data infrastructure necessary to construct genomic depend on models of paticnts, with the support of real-time methods for pattern recognition.
c) E-health will turn out a self-customizable medical solution, that will link cyber-physical to be integrated with big-data infrastructure and change the characteristics of implantable medical devices used to manage professional medical treatment (reconfiguring its nodes) in reply to the active restructuring of computer-based models, that can customize by human involvement learning.

This expertise will simplify the combination of patient-associated medical care with individualized treatment and will enable thousands of nodes to correlate with each other [49–58].

11.4 Research Materials and Methods

11.4.1 Dataset

The datasct WDBC "Wisconsin Diagnostic Breast Cancer" was used for the experimentation and is available at the UCI repository of Machine Learning [59]. This data was created by William Wolberg at the University of Wilconsin Hospitals. It consists of 699 samples with benign (65.5%) and malignant (34.5%).

11.4.2 Data Pre-Processing

Pre-processing includes two steps: one is discretization and the second is SMOTE. Discretization is defined as the method by which values are converted into discrete. SMOTE is used to balance the class. Liu *et al.* discussed a statical method known as discretization as Chi2 [60]. For class-balancing when the count of samples in the various category is not allocated evenly distributed. With the help of SMOTE, the imbalance dataset dilemma is addressed using oversampling. In SMOTE [61], to balance the different classes, samples are produced with the support of a k-nearest neighbor. The overall combination of discretizing and SMOTE helps in enhancing the performance of the results. After the pre-processing, the feature selection [62–64] technique is applied to enhance the performance of classifiers and reduced the time of computation. BFS-CFS [65] is used for feature selection in which for searching BFS (best first search) is used and for evaluator CFS (correlation-based feature selection) is used. It a wrapper selection of feature in accordance with supervised feature selection.

11.4.3 Classification Algorithms

The algorithm used for classification in the experiment serves various categories like Bayesian, trees, lazy and miscellaneous classifiers implemented in WEKA 3.9.4 [66].

Naïve Bayes [67] classifiers use probabilistic relationship among class and predicted variables. This classifier concludes the conditional class probability depending on the Bayes theorem and shows the simple distribution. Naïve Bayes classifier, that works well once classes in training datasets are imbalanced and it evaluates feature likelihoods for each class x depending on the complement of x, it means on rest of the samples of the datasets ignoring the training examples of class x itself. Deep Learning [68] perform classification and regression with the support of a multilayer perceptron. RBF classifier [69] construct a network with the help of k-means clustering process to run the basis function and learn linear or logistic regression. Sequential minimal optimization (SMO) [70] is a quadratic programming optimization algorithm proposed by C. Platt. It was designed for sparse data performance and linear SVM. It breaks large quadratic problems into a set of smaller quadratic problems, to minimize computation time. A decision table constructs a decision table by choosing the correct subset of features [71]. Part construct decision trees (C4.5) and extract identify directions from the constructed tree. As C4.5 uses an approach top-down, recursive divide and conquer which is considered as a standard classifier. AdaBoost M1 algorithm adjusts weights of training instance to enforce learning to strengthen the misclassified instances and it selects only those features which help in increasing the predictive power of the model [72]. Random Forest is a supervised classification algorithm, numerous numbers of decision trees are combined to form the random forest. Each tree includes a rule-based system. Like a decision tree, there is no need to calculate information gain in the case of random forest to find the root. The rule used to create the tree are used to predict the result [73]. Kernel logistic regression [74] is a method that expands the ordinary logistic regression to cope with information that is not linearly distinguishable. It requires kernel constraints and kernel function. Using RBF (radial basis function) as the kernel function. MLP [75] classifier is a class of ANN (feed-forward artificial neural network). It precisely refers to a network comprised of various layers along with an activation function. MLP described as vanilla when it has only a single hidden layer. LibSVM [76] operates faster than SMO because it uses *LibSVM* to develop the SVM classifier. The instance-based classifier is K*, in which the class of a test instance will be based upon the class of the ones that are training instances like it, as decided by some resemblance function. It is different from other instance-in in accordance with the learners in the sense that it uses an entropy-centered distance function [77]. The *classifier JCBA* works in conjunction with class association rules, which make up the association rules somewhere solely one class characteristic-value-pair is permitted in the outcome [78]. In the optimized forest, the merging of the genetic algorithm along with the random forest is done to enhance the results that have been accomplished only by the random forest classifier [67, 79–80].

11.5 Experimental Outcomes

The performance of various fifteen classifiers is shown in Tables 11.1 and 11.2. Table 11.1 shows the results before applying the feature selection. Table 11.2 shows the performance result after applying the feature selection method. Before feature selection, Bayes Net outshines showing precision of 97%, recall of 97.1% and MCC of 93.8%. But performance after applying the feature selection improves with each classifier shown in Table 11.2. Bayes Net, LibSVM, K-Star and RBF classifier are different classifier showing good results as 98% in precision, recall as 98% and MCC as 96%.

Table 11.1 Classification results of various machine learning algorithm on Wilcoxon breast cancer datasets.

ML classifiers	Precision	Recall	MCC
Bayes Net	**0.972**	**0.971**	**0.938**
DL	0.967	0.967	0.928
Kernel Logistic Regression	0.966	0.966	0.924
MLP Classifier	0.963	0.963	0.918
LibSVM	0.778	0.664	0.128
LibLinear	0.547	0.624	-0.001
SMO	0.967	0.967	0.927
K-Star	0.818	0.813	0.572
JCBA	0.908	0.908	0.797
J48	0.946	0.946	0.88
Optimized Forest	0.965	0.964	0.922
Random Forest	0.968	0.967	0.928
Decision Table	0.954	0.954	0.899
RBF Classifier	0.966	0.966	0.925
ADA BoostM1	0.948	0.948	0.886

With the benefit of feature selection, the performance of fifteen classifiers has improved. In the case of MCC, there is an increment of 2.2%. In the case of recall and precision, there is an increment of approximately 1%.

Comparison of fifteen classifiers with parameters precision, recall and MCC is shown through the graph in Figures 11.1 to 11.4.

Table 11.2 Classification results of various machine learning algorithm on Wilcoxon breast cancer datasets after feature selection.

ML classifiers	**Precision**	**Recall**	**MCC**
Bayes Net	**0.98**	**0.98**	**0.96**
DL	0.977	0.977	0.953
Kernel Logistic Regression	0.976	0.976	0.951
MLP Classifier	0.973	0.973	0.947
LibSVM	**0.98**	**0.98**	**0.96**
LibLinear	0.976	0.976	0.951
SMO	0.976	0.976	0.951
K-Star	**0.98**	**0.98**	**0.96**
JCBA	0.959	0.957	0.916
J48	0.957	0.957	0.915
Optimized Forest	0.97	0.97	0.94
Random Forest	0.967	0.967	0.934
Decision Table	0.962	0.962	0.924
RBF Classifier	**0.98**	**0.98**	**0.96**
ADA BoostM1	0.962	0.962	0.923

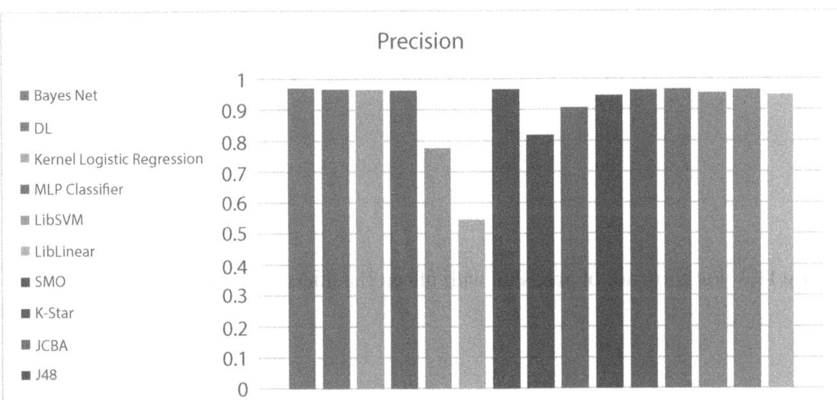

Figure 11.1 Performance of precision with fifteen classifiers without feature selection.

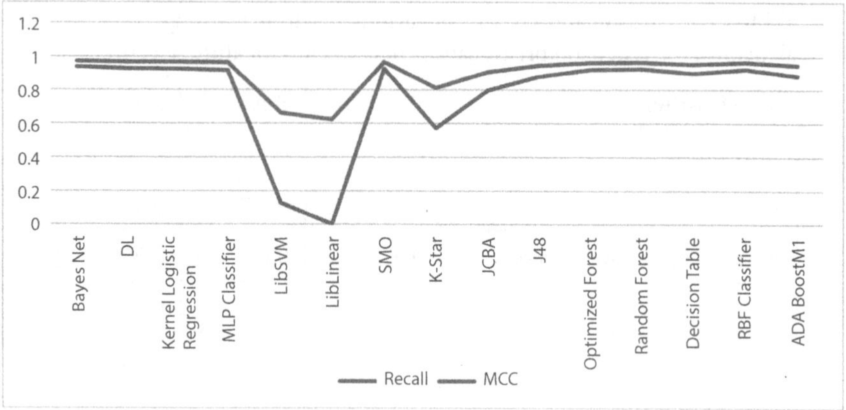

Figure 11.2 Performance of Recall and MCC with fifteen classifiers without feature selection.

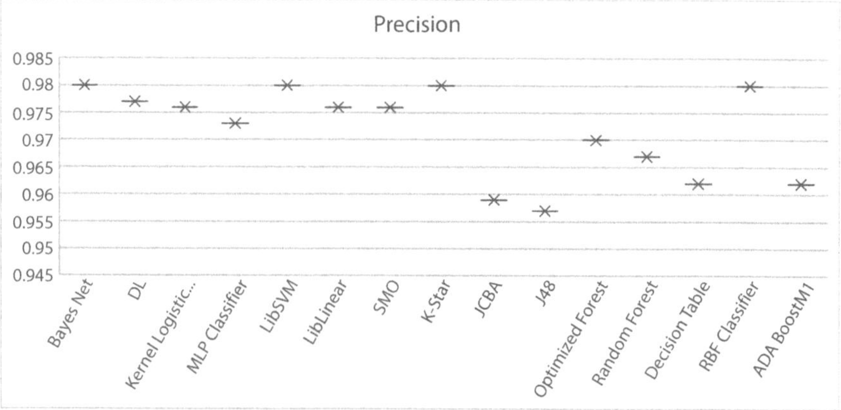

Figure 11.3 Performance of precision with fifteen classifiers with feature selection.

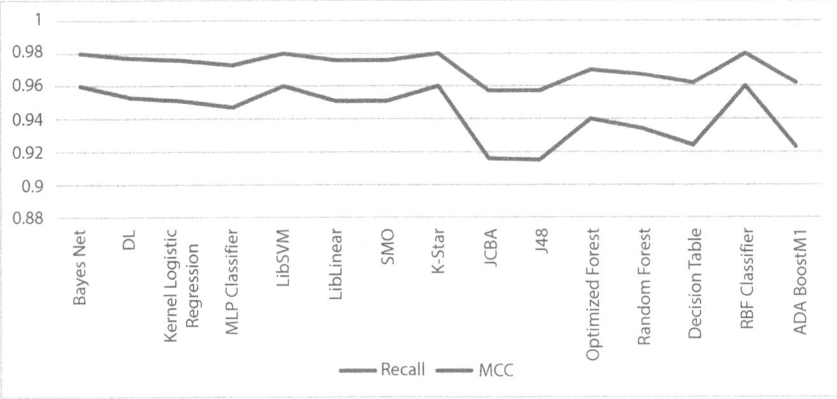

Figure 11.4 Performance of Recall and MCC with fifteen classifiers with feature selection.

11.6 Conclusion

Over the last few years, Internet of Things (IoT) has witnessed the transition process in a lifetime which offers the possibility to examine mutually both the real-time data and historical data by the changing role of data mining and artificial intelligence practices. In the early phase through the IoT the medical environment, the precise and efficient method for breast cancer diagnosis is particularly required for the treatment and recovery. In this research study, a couple of diagnostic system for accurate diagnosis of breast cancer are discussed in detail. In constructing the scheme for the machine learning acting like the forecasting model, many methods of machine learning were used for its discovery.

The existing state-of-the-art method does not successfully identify breast cancer in the initial stages, and most of the females have been suffering from this potentially hazardous disease. Consequently, the early discovery of breast cancer substantially poses a major challenge for healthcare professionals and scientists. To address the problem of early on-stage exposure of breast cancer, proposed a machine learning-centered diagnostic system that essentially organizes the benign and malignant individuals into the environment of IoT. In the foreseeable future, other features selection procedures, increasing the efficiency (optimization), and deep neural network categorization methods will be exploited to further boost the performance of the diagnostic procedure for BC diagnosis.

References

1. American Institute of Cancer Research, *Breast cancer statistics*, American Cancer Society, 2018, https://www.wcrf.org/dietandcancer/cancertrends/breast-cancer-statistics. https://www.cancer.org/content/dam/cancer-org/research/cancer-facts-and-statistics/annual-cancer-facts-and-figures/2018/cancer-facts-and-figures-2018.pdf
2. Islam, M.M., Iqbal, H., Haque, M.R., Hasan, M.K., Prediction of breast cancer using support vector machine and K-Nearest neighbors, in: *2017 IEEE Region 10 Humanitarian Technology Conference (R10-HTC)*, pp. 226–229, IEEE, 2017, December.
3. Ahmad, A.M., Khan, G.M., Mahmud, S.A., Miller, J.F., Breast cancer detection using cartesian genetic programming evolved artificial neural networks, in: *Proceedings of the 14th annual conference on Genetic and evolutionary computation*, pp. 1031–1038, 2012, July.
4. Azar, A.T. and El-Said, S.A., Probabilistic neural network for breast cancer classification. *Neural Comput. Appl.*, *23*, 6, 1737–1751, 2013.
5. Warner, E., Messersmith, H., Causer, P., Eisen, A., Shumak, R., Plewes, D., Systematic review: Using magnetic resonance imaging to screen women at high risk for breast cancer. *Ann. Intern. Med.*, *148*, 9, 671–679, 2008.
6. Haq, A.U., Li, J.P., Memon, M.H., Nazir, S., Sun, R., A hybrid intelligent system framework for the prediction of heart disease using machine learning algorithms. *Mob. Inf. Syst.*, 21, 2018.
7. Dey, A., Machine learning algorithms: A review. *Int. J. Comput. Sci. Inf. Technol.*, *7*, 3, 1174–1179, 2016.
8. Khan, M.A. and Salah, K., IoT security: Review, blockchain solutions, and open challenges. *Future Gener. Comput. Syst.*, *82*, 395–411, 2018.
9. Bhattacharya, S. and Pandey, M., Significance of IoT in India's E-Medical Framework: A study, in: *2020 First International Conference on Power, Control and Computing Technologies (ICPC2T)*, pp. 321–324, IEEE, 2020, January.
10. Kourou, K., Exarchos, T.P., Exarchos, K.P., Karamouzis, M.V., Fotiadis, D.I., Machine learning applications in cancer prognosis and prediction. *Comput. Struct. Biotechnol. J.*, *13*, 8–17, 2015.
11. Hameed, R.T., Mohamad, O.A., Hamid, O.T., Tapus, N., Design of e-Healthcare management system based on cloud and service oriented architecture, in: *2015 E-Health and Bioengineering Conference (EHB)*, pp. 1–4, IEEE, 2015, November.
12. Hsu, J.L., Hung, P.C., Lin, H.Y., Hsieh, C.H., Applying under-sampling techniques and cost-sensitive learning methods on risk assessment of breast cancer. *J. Med. Syst.*, *39*, 4, 1–13, 2015.
13. Parekh, M. and Saleena, B., Designing a cloud based framework for healthcare system and applying clustering techniques for region wise diagnosis. *Proc. Comput. Sci.*, *50*, 537–542, 2015.

14. Vijayarani, S. and Dhayanand, S., Data mining classification algorithms for kidney disease prediction. *Int. J. Cybern. Inform.*, *4*, 4, 13–25, 2015.
15. Turanoglu-Bekar, E., Ulutagay, G., Kantarcı-Savas, S., Classification of thyroid disease by using data mining models: A comparison of decision tree algorithms. *Oxford J. Intell. Decis. Data Sci.*, *2*, 13–28, 2016.
16. Verma, L., Srivastava, S., Negi, P.C., A hybrid data mining model to predict coronary artery disease cases using non-invasive clinical data. *J. Med. Syst.*, *40*, 7, 1–7, 2016.
17. Jahangir, M., Afzal, H., Ahmed, M., Khurshid, K., Nawaz, R., An expert system for diabetes prediction using auto tuned multi-layer perceptron, in: *2017 Intelligent Systems Conference (IntelliSys)*, pp. 722–728, IEEE, 2017, September.
18. Forkan, A.R.M., Khalil, I., Atiquzzaman, M., ViSiBiD: A learning model for early discovery and real-time prediction of severe clinical events using vital signs as big data. *Comput. Networks*, *113*, 244–257, 2017.
19. Osman, A.H. and Aljahdali, H.M., Diabetes disease diagnosis method based on feature extraction using K-SVM. *Int. J. Adv. Comput. Sci. Appl.*, *8*, 1, 2017.
20. Zhang, L., Zhou, W., Wang, B., Zhang, Z., Li, F., Applying 1-norm SVM with squared loss to gene selection for cancer classification. *Appl. Intell.*, *48*, 7, 1878–1890, 2018.
21. Kaur, P., Kumar, R., Kumar, M., A healthcare monitoring system using random forest and internet of things (IoT). *Multimed. Tools Appl.*, *78*, 14, 19905–19916, 2019.
22. Memon, M.H., Li, J.P., Haq, A.U., Memon, M.H., Zhou, W., Breast cancer detection in the IOT health environment using modified recursive feature selection. *Wireless Commun. Mobile Comput.*, 2019, 19, 2019.
23. Zhang, D., Zou, L., Zhou, X., He, F., Integrating feature selection and feature extraction methods with deep learning to predict clinical outcome of breast cancer. *IEEE Access*, *6*, 28936–28944, 2018.
24. Xiao, Y., Wu, J., Lin, Z., Zhao, X., Breast cancer diagnosis using an unsupervised feature extraction algorithm based on deep learning, in: *2018 37th Chinese Control Conference (CCC)*, pp. 9428–9433, IEEE, 2018, July.
25. Islam, M.M., Iqbal, H., Haque, M.R., Hasan, M.K., Prediction of breast cancer using support vector machine and K-Nearest neighbors, in: *2017 IEEE Region 10 Humanitarian Technology Conference (R10-HTC)*, pp. 226–229, IEEE, 2017, December.
26. Chaitanya, V., Killedar, S.M., Revankar, D., Pushpa, M.S., Recognition and Prediction of Breast Cancer using Supervised Diagnosis, in: *2019 4th International Conference on Recent Trends on Electronics, Information, Communication & Technology (RTEICT)*, pp. 1436–1441, IEEE, 2019, May.
27. Kavitha, M. and Krishna, P.V., IoT-Cloud-Based Healthcare System Framework to Detect Breast Abnormality, in: *Emerging Research in Data Engineering Systems and Computer Communications*, pp. 615–625, Springer, Singapore, 2020.

28. Azar, A.T. and El-Said, S.A., Probabilistic neural network for breast cancer classification. *Neural Comput. Appl.*, *23*, 6, 1737–1751, 2013.
29. Liu, Y.Q., Wang, C., Zhang, L., Decision tree based predictive models for breast cancer survivability on imbalanced data, in: *2009 3rd International Conference on Bioinformatics and Biomedical Engineering*, pp. 1–4, IEEE, 2009, June.
30. Ahmad, F., Isa, N.A.M., Hussain, Z., Sulaiman, S.N., A genetic algorithm-based multi-objective optimization of an artificial neural network classifier for breast cancer diagnosis. *Neural Comput. Appl.*, *23*, 5, 1427–1435, 2013.
31. Hasan, M.K., Islam, M.M., Hashem, M.M.A., Mathematical model development to detect breast cancer using multigene genetic programming, in: *2016 5th International Conference on Informatics, Electronics and Vision (ICIEV)*, pp. 574–579, IEEE, 2016, May.
32. Albrecht, A.A., Lappas, G., Vinterbo, S.A., Wong, C., Ohno-Machado, L., Two applications of the LSA machine, in: *Proceedings of the 9th International Conference on Neural Information Processing, 2002. ICONIP'02*, vol. 1, pp. 184–189, IEEE, 2002, November.
33. Pena-Reyes, C.A. and Sipper, M., A fuzzy-genetic approach to breast cancer diagnosis. *Artif. Intell. Med.*, *17*, 2, 131–155, 1999.
34. Akay, M.F., Support vector machines combined with feature selection for breast cancer diagnosis. *Expert Syst. Appl.*, *36*, 2, 3240–3247, 2009.
35. Zheng, B., Yoon, S.W., Lam, S.S., Breast cancer diagnosis based on feature extraction using a hybrid of K-means and support vector machine algorithms. *Expert Syst. Appl.*, *41*, 4, 1476–1482, 2014.
36. Ramadevi, G.N., Rani, K.U., Lavanya, D., Importance of feature extraction for classification of breast cancer datasets, a study. *Int. J. Sci. Innov. Math. Res.*, *3*, 2, 763–368, 2015.
37. Abbass, H.A., An evolutionary artificial neural networks approach for breast cancer diagnosis. *Artif. Intell. Med.*, *25*, 3, 265–281, 2002.
38. Marcano-Cedeño, A., Quintanilla-Domínguez, J., Andina, D., WBCD breast cancer database classification applying artificial metaplasticity neural network. *Expert Syst. Appl.*, *38*, 8, 9573–9579, 2011.
39. Liu, Y.Q., Wang, C., Zhang, L., Decision tree based predictive models for breast cancer survivability on imbalanced data, in: *2009 3rd International Conference on Bioinformatics and Biomedical Engineering*, pp. 1–4, IEEE, 2009, June.
40. Onan, A., A fuzzy-rough nearest neighbor classifier combined with consistency-based subset evaluation and instance selection for automated diagnosis of breast cancer. *Expert Syst. Appl.*, *42*, 20, 6844–6852, 2015.
41. Sheikhpour, R., Sarram, M.A., Sheikhpour, R., Particle swarm optimization for bandwidth determination and feature selection of kernel density estimation based classifiers in diagnosis of breast cancer. *Appl. Soft Comput.*, *40*, 113–131, 2016.

42. Rasti, R., Teshnehlab, M., Phung, S.L., Breast cancer diagnosis in DCE-MRI using mixture ensemble of convolutional neural networks. *Pattern Recognit.*, *72*, 381–390, 2017.
43. Ani, R., Krishna, S., Anju, N., Aslam, M.S., Deepa, O.S., Iot based patient monitoring and diagnostic prediction tool using ensemble classifier, in: *2017 International Conference on Advances in Computing, Communications and Informatics (ICACCI)*, pp. 1588–1593, IEEE, 2017, September.
44. Yang, Z., Zhou, Q., Lei, L., Zheng, K., Xiang, W., An IoT-cloud based wearable ECG monitoring system for smart healthcare. *J. Med. Syst.*, *40*, 12, 1–11, 2016.
45. Tsanas, A., Little, M.A., McSharry, P.E., Ramig, L.O., Nonlinear speech analysis algorithms mapped to a standard metric achieve clinically useful quantification of average Parkinson's disease symptom severity. *J. R. Soc. Interface*, *8*, 59, 842–855, 2011.
46. Cai, Z., Gu, J., Chen, H.L., A new hybrid intelligent framework for predicting Parkinson's disease. *IEEE Access*, *5*, 17188–17200, 2017.
47. Urbanowicz, R.J., Meeker, M., La Cava, W., Olson, R.S., Moore, J.H., Relief-based feature selection: Introduction and review. *J. Biomed. Inf.*, *85*, 189–203, 2018.
48. Naranjo, L., Perez, C.J., Martin, J., Campos-Roca, Y., A two-stage variable selection and classification approach for Parkinson's disease detection by using voice recording replications. *Comput. Methods Programs Biomed.*, *142*, 147–156, 2017.
49. Farahani, B., Firouzi, F., Chang, V., Badaroglu, M., Constant, N., Mankodiya, K., Towards fog-driven IoT eHealth: Promises and challenges of IoT in medicine and healthcare. *Future Gener. Comput. Syst.*, *78*, 659–676, 2018.
50. Gubbi, J., Buyya, R., Marusic, S., Palaniswami, M., Internet of Things (IoT): A vision, architectural elements, and future directions. *Future Gener. Comput. Syst.*, *29*, 7, 1645–1660, 2013.
51. Atzori, L., Iera, A., Morabito, G., The internet of things: A survey. *Comput. Networks*, *54*, 15, 2787–2805, 2010.
52. Da Xu, L., He, W., Li, S., Internet of things in industries: A survey. *IEEE Trans. Ind. Inf.*, *10*, 4, 2233–2243, 2014.
53. Zheng, Z., Zhu, J., Lyu, M.R., Service-generated big data and big data-as-a-service: an overview, in: *2013 IEEE international congress on Big Data*, pp. 403–410, IEEE, 2013, June.
54. Keim, D., Kohlhammer, J., Ellis, G., Mansmann, F., *Mastering the information age: solving problems with visual analytics*, 2010.
55. Uckelmann, D., Harrison, M., Michahelles, F., An architectural approach towards the future internet of things, in: *Architecting the Internet of Things*, pp. 1–24, Springer, Berlin, Heidelberg, 2011.
56. Miorandi, D., Sicari, S., De Pellegrini, F., Chlamtac, I., Internet of things: Vision, applications and research challenges. *Ad Hoc Networks*, *10*, 7, 1497–1516, 2012.

57. Sundmaeker, H., Guillemin, P., Friess, P., Woelfflé, S., Vision and challenges for realising the Internet of Things, in: *Cluster of European Research Projects on the Internet of Things, European Commision*, vol. 3(3), pp. 34–36, 2010.
58. Bandyopadhyay, D. and Sen, J., Internet of things: Applications and challenges in technology and standardization. *Wirel. Pers. Commun.*, *58*, 1, 49–69, 2011.
59. Wolberg, W.H., Street, N., Mangasarian, O.L., *Wisconsin diagnostic breast cancer (wdbc)*, U. o. California, Ed. (Ed.), University of Wisconsin, USA, 1995.
60. Liu, H. and Setiono, R., Chi2: Feature selection and discretization of numeric attributes, in: *Proceedings of 7th IEEE International Conference on Tools with Artificial Intelligence*, 1995, November, IEEE, pp. 388–391.
61. Jishan, S.T., Rashu, R.I., Haque, N., Rahman, R.M., Improving accuracy of students' final grade prediction model using optimal equal width binning and synthetic minority over-sampling technique. *Decis. Anal.*, *2*, 1, 1–25, 2015.
62. Lamba, M., Munjal, G., Gigras, Y., Feature Selection of Micro-array expression data (FSM)—A Review. *Proc. Comput. Sci.*, *132*, 1619–1625, 2018.
63. Lamba, M., Munjal, G., Gigras, Y., Computational Studies in Breast Cancer, in: *Diagnostic Applications of Health Intelligence and Surveillance Systems*, pp. 106–126, 2021.
64. Lamba, M., Munjal, G., Gigras, Y., Computational studies on breast cancer analysis. *J. Stat. Manage. Syst.*, *23*, 6, 999–1009, 2020.
65. Lamba, M., Munjal, G., Gigras, Y., A hybrid Gene Selection Model for Molecular Breast Cancer Classification using Deep Neural Network. *Int. J. Appl. Pattern Recognit.*, *6*, 3, 2021, In press. https://www.inderscience.com/info/ingeneral/forthcoming.php?jcode=ijapr
66. Hall, M., Frank, E., Holmes, G., Pfahringer, B., Reutemann, P., Witten, I.H., The WEKA data mining software: An update. *ACM SIGKDD Explor. Newsl.*, *11*, 1, 10–18, 2009.
67. Rish, I., An empirical study of the naive Bayes classifier, in: *IJCAI 2001 Workshop on Empirical Methods in Artificial Intelligence*, 2001, August, vol. 3, No. 22, pp. 41–46.
68. Goodfellow, I., Bengio, Y., Courville, A., Bengio, Y., *Deep Learning*, vol. 1, No. 2, MIT Press, Cambridge, 2016.
69. Gao, M., Hong, X., Chen, S., Harris, C.J., A combined SMOTE and PSO based RBF classifier for two-class imbalanced problems. *Neurocomputing*, *74*, 17, 3456–3466, 2011.
70. Zeng, Z.Q., Yu, H.B., Xu, H.R., Xie, Y.Q., Gao, J., Fast training support vector machines using parallel sequential minimal optimization, in: *2008 3rd International Conference on Intelligent System and Knowledge Engineering*, 2008, November, vol. 1, IEEE, pp. 997–1001.

71. Kumar, A. and Sarkar, B.K., A hybrid predictive model integrating C4. 5 and decision table classifiers for medical data sets. *J. Inf. Technol. Res. (JITR)*, *11*, 2, 150–167, 2018.
72. Eibl, G. and Pfeiffer, K.P., How to make AdaBoost. M1 work for weak base classifiers by changing only one line of the code, in: *European Conference on Machine Learning*, 2002, August, Springer, Berlin, Heidelberg, pp. 72–83.
73. Belgiu, M. and Drăguţ, L., Random forest in remote sensing: A review of applications and future directions. *ISPRS J. Photogramm. Remote Sens.*, *114*, 24–31, 2016.
74. Keerthi, S.S., Duan, K.B., Shevade, S.K., Poo, A.N., A fast dual algorithm for kernel logistic regression. *Mach. Learn.*, *61*, 1, 151–165, 2005.
75. Windeatt, T., Accuracy/diversity and ensemble MLP classifier design. *IEEE Trans. Neural Networks*, *17*, 5, 1194–1211, 2006.
76. Chang, C.C. and Lin, C.J., LIBSVM: A library for support vector machines. *ACM Trans. Intell. Syst. Technol. (TIST)*, *2*, 3, 1–27, 2011.
77. Mahmood, D.Y. and Hussein, M.A., Intrusion detection system based on K-star classifier and feature set reduction. *Int. Organ. Sci. Res. J. Comput. Eng. (IOSR-JCE)*, *15*, 5, 107–112, 2013.
78. Alaiad, A., Najadat, H., Al-Mnayyis, N., Khalil, A., Associative classification of the jordanian hospitals efficiency based on DEA, in: *2018 IEEE Conference on Big Data and Analytics (ICBDA)*, 2018, November, IEEE, pp. 23–26.
79. Kumar, N.K., Vigneswari, D., Krishna, M.V., Reddy, G.P., An optimized random forest classifier for diabetes mellitus, in: *Emerging Technologies in Data Mining and Information Security*, pp. 765–773, Springer, Singapore, 2019.
80. Shrivastava, G., Peng, S.L., Bansal, H., Sharma, K., Sharma, M. (Eds.), *New Age Analytics: Transforming the Internet Through Machine Learning, IoT, and Trust Modeling*, CRC Press, Palm Bay, Florida, USA, 2020.

12

Perspective-Based Studies of Trust in IoHT and Machine Learning-Brain Cancer

Sweta Kumari[1*], Akhilesh Kumar Sharma[1†], Sandeep Chaurasia[1‡] and Shamik Tiwari[2$]

[1]*Manipal University Jaipur, Jaipur, India*
[2]*UPES, Dehradun, India*

Abstract
Healthcare Sector is the core of a nation's development, and a fit body can make great contributions towards the development of a nation. Since taking care of the brain will almost result in maintaining the health of the human body, so this chapter will focus on one of the important concerns in the healthcare sector i.e., brain tumor. The brain tumor is deadly for everyone, but it specifically affects children or patients with genetic neurological disease. The symptoms of the brain tumor vary as per the physiques of a patient which causes the delay in the diagnosis of the tumor. The fatality of these brain illnesses depends on the grades of brain cancer and some of the other factors which will be further discussed in this chapter along with these grades of a brain tumor in detail. Furthermore, there will be a discussion about various challenges that may arise during the classification of types of brain tumor. There are many automatic or semi-automatic conducts for sleuthing brain tumor, so the focus will be on the best methods in each mode for segmentation and detection of brain tumor. Treatment of brain cancer does not lead to the prolonged existence of patients. This chapter will also provide the survival rate of patients and these survival rates will be based on the different types of tumor. This chapter is an attempt to summarize all necessary information which needs to be further explored for a brain tumor as well as trust in IoHT and machine learning regarding brain cancer.

**Corresponding author*: swetakumari3031@gmail.com
†*Corresponding author*: akhileshshm@gmail.com
‡*Corresponding author*: chaurasia.sandeep@gmail.com
$*Corresponding author*: shamiktiwar@gmail.com

Kavita Sharma, Yogita Gigras, Vishnu Sharma, D. Jude Hemanth and Ramesh Chandra Poonia (eds.) Internet of Healthcare Things: Machine Learning for Security and Privacy, (265–280) © 2022 Scrivener Publishing LLC

Keywords: Brain cancer, healthcare sector, IoHT, machine learning, brain tumor sleuthing

12.1 Introduction

Internet of Things (IoT) aids in real time data communication to the designated place through some of its devices. Internet of Healthcare Things (IoHT) and the use of machine learning has been considered further advanced technology for patients because of reasons such as effortlessness disease detection, reduction in demise rates, reduction in medical cost, etc. Some of the IoT devices are displays, movement tracker, camera, detectors, etc. Machine Learning uses some of its appropriate algorithms to reduce the errors made by the experts during decision making. These devices are being ingested/wearable into the patient's body as per the guidelines of the doctors. Figure 12.1 illustrates the process flow of IoHT and the use of machine learning in it.

Trust in IoHT and machine learning for brain cancer can be understood through these four steps:

- Data Acquisition: Clinical data acquisition from hospitals using IoT devices that are recommended by doctors as per patients' tumor conditions.
- Data Cleaning: Data tuning by removal of unwanted data that does not have any relation to the patient's tumor treatments.

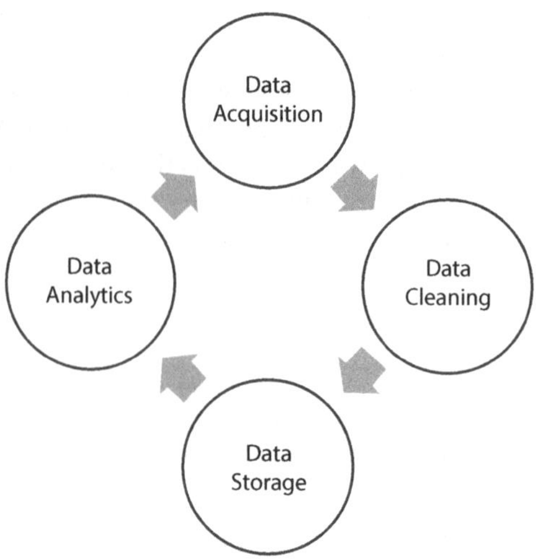

Figure 12.1 Process flow for internet in healthcare system.

- Data Storage: Tuned data storage on the cloud for access by experts or researchers.
- Data Analytics: Use of the data from the cloud to make decisions using machine learning as data analytics to reduce the error and time-lapse.

This chapter presents an illustration of trust in IoHT and machine learning for brain cancer. The remaining chapter is systematized as follows: Section 12.2 indicates the literature survey in association with the use of IoT and machine learning in brain cancer. Similarly, Section 12.3 gives the details about brain tumor. Section 12.4 provides the detail regarding the sleuthing of brain tumor and the challenges during the classification of tumor. Section 12.5 provides the survival rate for the tumor patients, followed by the conclusion in Section 12.6.

This chapter is organized to discuss literature survey and types of brain cancer and their associated challenges. The survival rate of various kinds of cancerous diseases is also discussed using some case studies.

12.2 Literature Survey

This section focuses on those works which are related to detection and segmentation of brain tumor with the use of IoT technology as well machine learning to build the trust on IoHT [15]. This paper had used IoT technology (Thingspeak) and a machine learning algorithm (Support Vector Machine (SVM)) to aid in the decision making of surgery through detection and segmentation of brain tumor. The result was better than the use of this algorithm without IoT [16]. This paper had used an IoT and cloud-based approach for the detection and classification of brain tumor. Magnetic resonance images as input were captured by the IoT devices which later sent through the cloud for the diagnosis of tumor. The optimal Dense Convolutional Network model was used to work on those collected MRI images which were received through the cloud and this model performed with 99.37% accuracy.

In this paper [17], Parzen Analysis of Brain Tissue Densities was used as an extractor and Support Vector Machine (SVM) was used as a classifier to determine the stroke and its type. IoT was used for storage of configuration material of Parzen Analysis of Brain Tissue Densities along with its classifier details on the web and then for extraction and prediction on the CT scans of the brain. An accuracy of 98.41% was achieved for this proposed method which was better than other methodologies. Reference [18] focused on the use of Internet of Medical Things (IoMT) as well as the Deep Convolution Neural Network (D-CNN) model for the classification of brain tumor. Data from brain scans using MRI was collected and send through IoT devices. On the cloud, these data as well as well-trained model's data were stored for free access from any location. Deep Convolution Neural Network model was used for recognition of tumor area and classification of it into its types. An accuracy of 98.07% was achieved for this proposed method which was better compared to other methodologies.

Reference [19] focused on the computation of fractal model to provide aid to the doctors specializing in medical imaging. This IoT and cloud-based model was used for the procedure of brain tumor images with the help of different methods of the fractal model. These analyzed results of brain images were communicated using IoT-cloud based storage that saved manpower as well as time and improved accuracy. Reference [20] focused on the use of medical devices and IoT for communication of patient's medical brain images. This paper was further divided into a selection of features as well as classification of brain tumor. The MRI images received through IoT was further analyzed and then an improved gravitational search algorithm with a genetic algorithm was used for feature mining. Support Vector Machine (SVM) was used as a classifier to classify brain tumor into their types. This model performed better than other models.

Reference [21] focused on the prediction of disease using wearable IoT devices. The focus of this paper was to use deep brain stimulation as a collector of brain-related data and a heuristic tubu optimized sequence modular neural network for processing this data. This examined data was helpful in the prediction of brain diseases in an independent way with better accuracy. This method performed better for monitoring changes in brain function.

12.3 Illustration of Brain Cancer

This section provides an illustration of brain cancer along with its types as well as grades of tumor. It also exhibits some of the symptoms that may help in the diagnosis of the tumor.

12.3.1 Brain Tumor

When the genetic DNA of the damaged cells gathers at a place then it is shaped into a brain tumor. An estimated eighty-seven thousand individuals in the US with non-cancerous tumors was diagnosed in 2020 [1]. It can be assumed that the ratio can be 60:20 for non-cancerous and cancerous tumors. For both male and females, the 10th most lethal reason for death is cancer associated with brain and another nervous system [2]. The estimation for death from a primary cancerous brain tumor and CNS tumor in 2020 consists of 10,190 men and 7,830 women (18,020 adults) [2]. A vigorous human brain comprised of ordinary brain matters but when this ordinary brain tissue shows some anomalous tissues along with the ordinary tissues then these anomalous tissues are baptized as a brain tumor and tumor's name is given grounded on the tissue it originates from or on the origin of their location in the brain. A brain tumor can be homogeneous or heterogeneous depending on the kinds of tumor. The visual of tumor varies based on its intensity level. The intensity level varies depending on the type of tumor. If a patient has a benign tumor with the high-level intensity that means tumor's section seems brighter than the immediate area while a patient with a low-level intensity tumor means the tumor's section seems darker than the immediate area, but the borderline seems to be brighter than the tumor's section.

12.3.2 Types of Brain Tumors

Several diverse kinds of brain cancer are available which can be approximately over 120 tumors, but brain tumor can be mainly said to be of two types such as primary brain tumor (non-cancerous or benign), and secondary brain tumor (cancerous or malignant).

- **Primary Brain Tumor:** Primary brain tumor has their own borderline to grow gradually, and these restrictions cover only brain parts but not the additional portions of the body. Generally, for a primary tumor the signal intensity level remains undeviating. In 2020, the estimation of total patients to be diagnosed from primary cancerous tumors of the brain and spinal cord in the United States are 13,590 men and 10,300 women (23,890 adults) [2]. Some of the benign tumors are Meningiomas, Schwannoma, and Gliomas with its types such as astrocytoma (Grades I to IV), and Oligodendroglioma, etc.
- **Malignant Brain Tumor:** Secondary brain tumor (Malignant) does not follow any borderline to obey with which makes these types of tumor more complex as well as deadly. This type of tumor grows faster than non-cancerous tumors. In the case of this type of tumor, the signal intensity level continues deviating as it might start from diverse location and reach to brain. In 2020, the estimation of 3,700 cases

in adolescents and young adults were expected to be diagnosed for malignant brain tumors [3]. Some of the malignant tumors that invade parts of the brain are lung cancer (metastatic lung cancer), breast cancer, kidney cancer, etc.

12.3.3 Grades of Brain Tumors

As brain tumor is classified into two types i.e., primary brain tumor which can also be known as low-grade tumor (Benign) and secondary brain tumor (Malignant) known as a high-grade tumor. The low-grade tumor is known to be less lethalness in contrast to a high grade tumor. These types of brain tumor are further alienated into grades of tumor which ranges from grade I to grade IV. This grading of tumor was done by World Health Organization to further comprehend the risk level and provide an apposite type of treatment to the patients. These different grades of tumor are characterized according to the structure of tumor perceptible through the microscope. Grades I and II are considered primary brain tumor while grades III and IV are considered secondary brain tumor. As previously it was distinguished that the benign tumor was less deadly compared to the Malignant tumor accordingly grades I and II become less deadly compared to grades III and IV. Figure 12.2 exhibits the different grades of tumor.

Discussion of each grade of tumor along with some names of the tumor are stated below:

- **Grade I tumor:** A grade I tumor can be effortlessly removed as it seldom invades from the brain to other parts of the human body and it has an advantage over other types of tumor that it grows at a slow rate. Gangliocytoma, Pilocytic astrocytoma, Ganglioglioma, and Craniopharyngioma are some of the tumors that fall under the category of this grade of tumor.

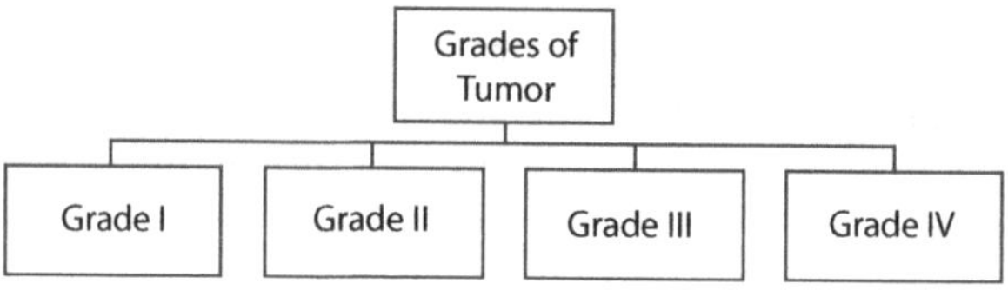

Figure 12.2 Grades of brain tumor.

- **Grade II tumor:** This grade of tumor is somewhat diverse as it does look slightly different than usual brain tissues and may spread to other parts of the body which may intensify its level of deadliness. Pineocytoma, "Diffuse" Astrocytoma, and Pure oligodendroglioma are a few examples of this grade of tumor.
- **Grade III tumor:** This grade of tumor grows faster than the above-mentioned grades so it's difficult to provide the appropriate treatment and if treated may reoccur. This grade of tumor spreads to other parts of body at an extremely high speed which if not treated at the precise time may lead to death. Anaplastic oligodendroglioma, Anaplastic ependymoma, and Anaplastic astrocytoma are some of these grades of tumor.
- **Grade IV tumor:** To keep up quick development grade IV tumor shapes fresh blood vessels. This grade of tumor is the most deadliness tumor which has dead cells and it is difficult to eradicate because eradication may lead to the elimination of some of the normal working cells. This type of tumor may not be detached only with the surgery but may also prerequisite additional treatment as it quickly grows back to other parts of the body. Sometimes there may be conditions where both low grade and the high-grade tumor may be found in a patient, then this type of tumor is categorized based on the type of tumor which prevails more. Medulloblastoma, Gliobastoma multiforme, Ependymoblastoma, and Pineoblastoma are some of grade IV tumor.

12.3.4 Symptoms of Brain Tumors

The brain tumor is different from other types of tumor as patients having different types of brain tumor may demonstrate similar symptoms but the grade of tumor for patients may diverge which makes detection of brain tumor very problematic. Symptoms of a brain tumor may also vary due to the physique of patients as well the size of tumor. The patients have different body structure as well as different immune system leading to alteration in symptoms of cancer. Figure 12.3 exhibits some communal symptoms of cancer.

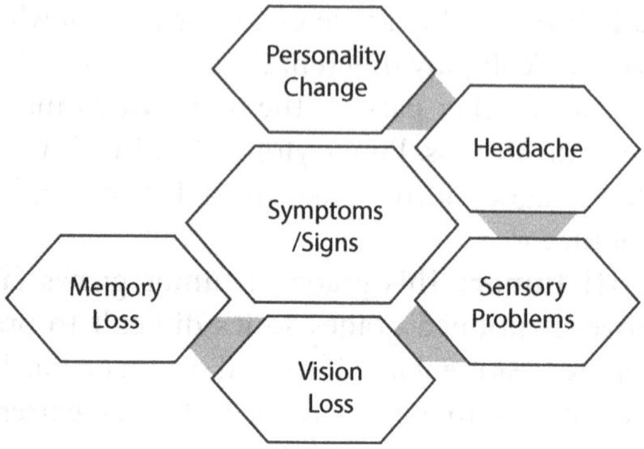

Figure 12.3 Common signs of cancer.

Some symptoms that are dependent on the position of brain cancer are as follows:

- **Frontal Lobe symptoms:** The symptoms such as variation in the personality leading to annoyance, exertion in way of talking as well as struggle in walking, and unresponsiveness on any side of the body which may cause loss of aroma are some of the symptoms related to tumors located on the frontal lobe of the brain.
- **Temporal Lobe symptoms:** Understanding problem leading to even misperception about ordinariness, absentmindedness which causes loss of concentration power to learn any new information or recognizing any object are some of the symptoms of brain tumor related to temporal lobe location of brain tumor.
- **Parietal Lobe symptoms:** Symptoms associated with tumors in the parietal lobe are one way or another similar to the temporal lobe in the case of recognition of substances, but it also has supplementary symptoms like both eye and hand have a problem in synchronization with each other, Sensory problems like hear, touch, smell, sight, taste also struggle in retorting to any kind of agony or any interior wound.
- **Occipital Lobe symptoms:** The symptoms such as vision loss either fully or partly, phantasm is some of the symptoms related to tumors located on the occipital lobe.
- **Brainstem symptoms:** Tumors located in the brainstem has symptoms which have a combination of almost all the

above symptoms such as variation in the personality leading to annoyance, Sensory problems like hear and sight, both eye and hand have a problem in synchronization with each other as well as other symptoms like exertion in swallowing, tiredness etc.

- **Cerebellum symptoms:** Tumors located in Cerebellum has a vision-related problem such as giddiness, obstreperous eye movement and muscle-related problems such as movement of arms, legs or any other part of the body, etc.

12.4 Sleuthing and Classification of Brain Tumors

This section describes the sleuthing of tumor and some of the best available methodologies based on different grounds. It also explicates some challenges that may hinder the classification of tumor.

12.4.1 Sleuthing of Brain Tumors

Brain tumor segmentation and detection's method contrasts and is dependent on numerous aspects, some of those aspects are a type of tumor, the intensity of tumor, location and size of tumor, time and cost, sequences of images used in dataset as well as several types of method preferred such as generative or discriminative method, instance, or semantic methods, automatic, semi-automatic, or manual methods. As various methods differ so does the best method for segmentation and detection might differ for individual's dependent on the above stated aspects.

In general, the discriminative method works better in comparison to the generative method because the former depends on the features mined from brain tumor images and the latter depends on the preceding knowledge of the patient's health and oncology issue.

Among methods such for instance or semantic segmentation, semantic segmentation seems to be better than instance segmentation because the semantic method can provide similar colors to the tumor region which belong to a precise class while the instance method provides separate colors to the tumor region belonging to a precise class which makes the work little difficult for the researchers. Some of the best semantic segmentation includes an encoder (or contracting path) and decoder (or an expansive path) [7, 8].

Among automatic, semi-automatic, or manual segmentation methods, the performance of each method varies case to case basis and on some of the aspects stated above. For the convenience of individuals, the fully automatic and semi-automatic method seems to better in comparison to manual segmentation methods because this method is very time consuming and costly as it requires an expert and this expert's decision may not be accurate. As for fully automatic or semi-automatic segmentation methods, the performance level depends on the type of result the researchers want.

In general, a fully automatic segmentation method can be stated better because neither prerequisite any expert for assigning segmented borderline nor user interaction is obligatory. Some of the best fully automatic-method for brain tumor can include deep CNN (Convolutional Neural Network) and its different models such as ResNet (Residual Neural Network) [7, 9], VGGNet (Visual Geometry Group Network) [7, 10], DenseNet (Dense Convolutional Network) [7, 11], Xception [7, 12], MobileNet [7, 13], and MobileNetV2 [7, 14].

12.4.2 Challenges During Classification of Brain Tumors

Identification of brain tumor accompanied by its type and grade of tumor is the decisive and also most challenging period. Some of those challenges which might affect the classification of brain tumor are as follows:

- Recognition of correct symptoms of brain tumor as well as its grade in a patient is the most challenging task as a recommendation of treatment or further medical exam to verify tumor from non-tumor depends on this recognition.
- After recognition, another crucial task that needs accuracy is analysis of tumor using small portion of tumor as a sample which is further treated as a block to classify tumor. The selection of sample is vital as the entire study on brain tumor depends on the sample provided to it. If the selection of sample is not balanced, then analysis of brain tumor becomes impractical.
- For an analysis to be done during neurosurgeons on a tumor's sample is challenging as the gathering of samples is problematic. As it is known, the doctor can't disclose a patient's medical details and so it affects the number of samples collected as well as there might be some legal complications.
- Making classification of tumor fully automatic based on features selected is a challenging task.

12.5 Survival Rate of Brain Tumors

The survival of a patient is reliant on the nature of treatment provided to them and the treatments can be surgery, chemotherapy, radiation, therapy, drugs, or a combination of some of these treatments. The survival rate for 5 years means the percent of people who remain alive for 5 years after treatment of tumor is completed. This percent of people survived is calculated out of 100 persons while for relative 5-year survival rate, the comparison is conducted between the number of people with similar kinds of tumor and the entire population.

For both types of brain tumors, the percentage for the overall tumor cases in adolescents and young adults (AYAs) has its range i.e., 48% in the case of adolescents and 63% in the case of young adults of age 30–39 years [3].

Benign and borderline brain tumor has approximately 97% relative survival rate of 5-years [3, 5]. This survival rate is higher altogether in all age groups of adolescents and young adults (AYAs) [3, 5].

The survival rate depends on numerous aspects other than just treatments provided to the patient, some of these aspects upon which survival rate depends are given below:

- The survival rate depends mainly on the kind of tumor a patient has. The severeness of the patient's condition directly depends on tumor type as benign tumor's level of severeness is diverse from a malignant tumor.
- The survival rate for brain tumor varies depending upon the grade of tumor. It can be stated that low-grade tumor has higher chance of survival compared to higher-grade tumor.
- The survival rate depends on the where tumor's position is situated in brain as early detection of tumor is hindered by a protective layer termed a skull. Tumor's location is vital, but the size of tumor also has a crucial part that may hamper the detection or appropriate treatment to be provided to the patient. For example, if the tumor is situated at a place where the skull hinders in the detection of this tumor but if the size of tumor is bulky it becomes easy to detect tumor even at that part of brain and vice versa.
- The physique of a patient also plays an imperative part in the survival rate. Patients having genetic health issues may not only show its effect on the early detection of tumor but also on the survival rate. Patients with a genetic health issue will show some symptoms which might be considered related to their respective genetic disease and this symptom can result in neglection of the brain tumor symptoms consequently primary detection of tumor can be easily affected.

- The age of the patient can also be one of the aspects upon which survival rate may depend. The age of a patient is causally related to the immune system or capability to hold off any reactions to the treatment given to patients. Suppose if the 5-year survival rate is 36% for malignant tumor then this percentage decreases to 31% when it comes to a 10-year survival rate [2].
- Treatment also someway plays a significant role in the survival of a patient as the physique of patients differ so does their type of treatment to be provided to them on a case-to-case basis should vary. Every human body does not retort to the treatments in the same way so there may be likelihoods that some patient may have an undesirable effect on their body and may not be able to survive the treatments.
- The accomplishment of surgery is the foremost key to survive the brain tumor.
- One more factor that is responsible for survival is the tumor's response to the treatment to cure the tumor due to intricacy in the tumor's molecular form.

A 5-year period survival rate based on gender is given below:

- The 5-year survival rate for the year 2016 for males were 30.1% [5].
- The 5 years survival rate for the year 2016 for females were 33.9% [5].

A 5-year survival rate (in the US, 2012–2016) based on types of brain tumor is given below:

- For patients of age 0–14 years, the 5-year survival rate for benign tumor and other CNS tumor was 74.7%, for patients of age 15–39 years was 71.3%, and patients of age more than 40 years had 21.3% 5-year survival rate [3, 4, 6]. AYAs patients of age between 15 and 19 years had 77% 5-year survival rate while 66% for patients of age 30–39 years [3, 4].
- For patients of age 0–14 years, the 5-year survival rate for benign tumor and other CNS tumor was 97.2%, for patients of age 15–39 years was 98.1%, and patients of age more than 40 years has 90.1% 5-year survival rate [6].

The 5-year relative survival rate for brain tumor as per the type of tumor for benign/borderline malignant tumor in United states is given below:

- For Age between birth and 14 years, the case distribution for year 2012 through 2016 was 8% and 5-year (2009 through 2015) survival rate was 97% [2].
- For Age between 15 and 19 years, the case distribution for year 2012 through 2016 was 13% and 5-year (2009 through 2015) survival rate was 98% [2].

Incidence rate, death rate, and relative survival rate for brain and another nervous system as per age (per 100,000 people) without primary brain tumor and borderline secondary brain tumors is given below:

- For Age between 15 and 19 years, the incidence rate for year 2012–2016 was 2.2, death rate for year 2013–2017 was 0.5, and 5-year (2009–2015) survival rate was 77% [3].
- For Age between 20 and 29 years, the incidence rate for year 2012-2016 was 2.5, death rate for year 2013–2017 was 0.6, and 5-year (2009–2015) survival rate was 73% [3].
- For Age between 30 and 39 years, the incidence rate for year 2012–2016 was 3.7, death rate for year 2013–2017 was 1.5, and 5-year (2009–2015) survival rate was 66% [3].

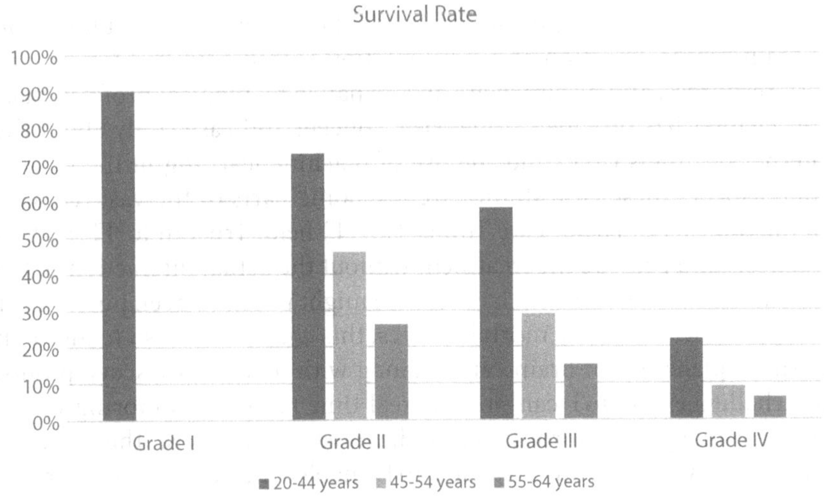

Figure 12.4 Survival rate for different grades of astrocytoma.

The survival rate for different grades of tumor varies depending on the different tumor. Figure 12.4 exhibits astrocytoma's survival rate in the US in the year 2001–2015 after diagnosis based on grade of tumor.

- Grade I tumor has a long-term survival rate if treated properly and it does not recur once treated. A 5-year survival rate for pilocytic astrocytoma (Grade 1) was greater than 90% [6].
- Grade II tumor sometimes recurs and once recurring it may be in the form of a high-grade tumor. In case of diffuse astrocytoma, patients of age 20–44 years had 73% 5-year survival rate, 46% for patients of age 45–54 years, and 26% for patients of age 55–64 years. [6].
- In case of anaplastic astrocytoma, patients of age 20–44 years had 58% 5-year survival rate, 29% for patients of age 45–54 years, and 15% for patients of age 55–64 years [6].
- In case of glioblastoma astrocytoma, patients of age 20–44 years had 22% 5-year survival rate, 9% for patients of age 45–54 years, and 6% for patients of age 55–64 years [6].

12.6 Conclusion

The healthcare sector is the most prominent industry which needed many innovations as well as smartness in its technology, but innovation should be accompanied by responsibility. In the healthcare sector, automation of brain tumor detection and segmentation within a limited timeframe, cost, resources, and accuracy are always hailed by medical professionals. This chapter has provided some basic concept and case study about the internet of things (IoT) and the use of machine learning in the field of brain cancer. These case studies are just some parts of the vast research on the Internet of Healthcare Things (IoHT) field. Trust in IoHT enables diagnosis and provide the treatment without the actual interaction of the patient with the doctor which otherwise might not have been possible for every patient. To reduce mortality rates, this chapter has also focused on a brief explanation of brain cancer along with its categories, symptoms, and challenges. As IoT can provide real time remote monitoring which can save many lives through early diagnosis of diseases. Therefore, in future, many novel and advanced IoT enabled devices as well various machine learning techniques can aid in the better diagnosis and treatment of tumor patients. Security of these patients' records used through these IoT devices as well as security of the cloud storage can also be covered in future work.

References

1. Mariotto, A.B., Robin Yabroff, K., Shao, Y., Feuer, E.J., Brown, M.L., Projections of the cost of cancer care in the United States: 2010–2020. *J. Natl. Canc. Inst.*, *103*, 2, 117–128, 2011.
2. Siegel, R.L., Miller, K.D., Goding Sauer, A., Fedewa, S.A., Butterly, L.F., Anderson, J.C., Cercek, A., Smith, R.A., Jemal, A., Colorectal cancer statistics, 2020. *CA: Cancer J. Clin.*, *70*, 3, 145–164, 2020.
3. Miller, K.D., Fidler-Benaoudia, M., Keegan, T.H., Hipp, H.S., Jemal, A., Siegel, R.L., Cancer statistics for adolescents and young adults, 2020. *CA: Cancer J. Clin.*, *70*, 6, 443–459, 2020.
4. Davies, L. and Welch, H.G., Current thyroid cancer trends in the United States. *JAMA Otolaryngol.—Head Neck Surg.*, *140*, 4, 317–322, 2014.
5. Howlader, N.N.A., Krapcho, M., Miller, D., Bishop, K., Kosary, C.L., Yu, M., Cronin, K.A. (Eds.), *SEER Cancer Statistics Review, 1975–2014*, National Cancer Institute, Bethesda, MD, 2017.
6. Ostrom, Q.T., Cioffi, G., Gittleman, H., Patil, N., Waite, K., Kruchko, C., Barnholtz-Sloan, J.S., CBTRUS statistical report: primary brain and other central nervous system tumors diagnosed in the United States in 2012–2016. *Neuro-oncology*, *21*, Supplement_5, v1–v100, 2019.
7. Zeineldin, R.A., Karar, M.E., Coburger, J., Wirtz, C.R., Burgert, O., DeepSeg: deep neural network framework for automatic brain tumor segmentation using magnetic resonance FLAIR images. *Int. J. Comput. Assist. Radiol. Surg.*, *15*, 6, 909–920, 2020.
8. Ronneberger, O., Fischer, P., Brox, T., U-net: Convolutional networks for biomedical image segmentation, in: *International Conference on Medical Image Computing and Computer-Assisted Intervention*, 2015, October, Springer, Cham, pp. 234–241.
9. He, K., Zhang, X., Ren, S., Sun, J., Deep residual learning for image recognition, in: *Proceedings of the IEEE Conference on Computer Vision and Pattern Recognition*, pp. 770–778, 2016.
10. Simonyan, K. and Zisserman, A., Very deep convolutional networks for large-scale image recognition. *Comput. Vis. Pattern Recognit.*, 14, 1–9, 2014.
11. Huang, G., Liu, Z., Van Der Maaten, L., Weinberger, K.Q., Densely connected convolutional networks, in: *Proceedings of the IEEE Conference on Computer Vision and Pattern Recognition*, pp. 4700–4708, 2017.

12. Chollet, F., Xception: Deep learning with depthwise separable convolutions, in: *Proceedings of the IEEE Conference on Computer Vision and Pattern Recognition*, pp. 1251–1258, 2017.
13. Howard, A.G., Zhu, M., Chen, B., Kalenichenko, D., Wang, W., Weyand, T., Andreetto, M., Adam, H., Mobilenets: Efficient convolutional neural networks for mobile vision applications. *Comput. Vis. Pattern Recognit.*, 17, 55–61, 2017.
14. Sandler, M., Howard, A., Zhu, M., Zhmoginov, A., Chen, L.C., Mobilenetv2: Inverted residuals and linear bottlenecks, in: *Proceedings of the IEEE Conference on Computer Vision and Pattern Recognition*, pp. 4510–4520, 2018.
15. Shrivastava, G., Peng, S.L., Bansal, H., Sharma, K., Sharma, M. (Eds.), *New age analytics: Transforming the internet through machine learning, IoT, and trust modeling*, CRC Press & Apple Academic Press, Florida, USA, 2020.
16. Ganesan, M., Sivakumar, N., Thirumaran, M., Saravanan, R., IOT And Cloud Based Brain Tumor Detection and Classification Model Using Optimal Densely Connected Convolutional Networks. *IIOAB J.*, 11, 5, 1–7, 2020
17. Sarmento, R.M., Vasconcelos, F.F., Rebouças Filho, P.P., de Albuquerque, V.H.C., An IoT platform for the analysis of brain CT images based on Parzen analysis. *Future Gener. Comput. Syst.*, *105*, 135–147, 2020.
18. Ganesan, M., Sivakumar, N., Thirumaran, M., Internet of medical things with cloud-based e-health services for brain tumour detection model using deep convolution neural network. *Electron. Gov., Int. J.*, *16*, 1–2, 69–83, 2020.
19. Lenka, S., Kumar, S., Mishra, S., Jena, K.K., An IoT-Cloud Based Fractal Model for Brain Tumor Image Analysis, in: *2020 Fourth International Conference on I-SMAC (IoT in Social, Mobile, Analytics and Cloud)(I-SMAC)*, 2020, October, IEEE, pp. 1–7.
20. Chidambaranathan, S., Radhika, A., Priya, V.V., Mohan, S.K., Gireeshan, M.G., Optimal SVM Based Brain Tumor MRI Image Classification in Cloud Internet of Medical Things, in: *Cognitive Internet of Medical Things for Smart Healthcare*, pp. 87–103, Springer, Cham, 2021.
21. AlZubi, A.A., Alarifi, A., Al-Maitah, M., Deep brain simulation wearable IoT sensor device based Parkinson brain disorder detection using heuristic tubu optimized sequence modular neural network. *Measurement*, *161*, 107887, 2020.

Index

Printed in the USA/Agawam, MA
September 16, 2026

910661.043